A TREATMENT IMPROVEMENT PROTOCOL

Managing Chronic Pain in Adults With or in Recovery From Substance Use Disorders

TIP 54

U.S. DEPARTMENT OF HEALTH AND HUMAN SERVICES
Substance Abuse and Mental Health Services Administration
Center for Substance Abuse Treatment

1 Choke Cherry Road
Rockville, MD 20857

Acknowledgments

This publication was prepared for the Substance Abuse and Mental Health Services Administration (SAMHSA) by the Knowledge Application Program (KAP), a Joint Venture of The CDM Group, Inc., and JBS International, Inc., under contract numbers 270-04-7049 and 270-09-0307, with SAMHSA, U.S. Department of Health and Human Services (HHS). Christina Currier served as the Government Project Officer.

Disclaimer

The views, opinions, and content of this publication are those of the authors and do not necessarily reflect the views, opinions, or policies of SAMHSA or HHS.

Public Domain Notice

All materials appearing in this volume except those taken directly from copyrighted sources are in the public domain and may be reproduced or copied without permission from SAMHSA or the authors. Citation of the source is appreciated. However, this publication may not be reproduced or distributed for a fee without the specific, written authorization of the Office of Communications, SAMHSA, HHS.

Electronic Access and Copies of Publication

This publication may be ordered from SAMHSA's Publications Ordering Web page at http://www.store.samhsa.gov. Or, please call SAMHSA at 1-877-SAMHSA-7 (1-877-726-4727) (English and Español). The document can be downloaded from the KAP Web site at http://www.kap.samhsa.gov.

Recommended Citation

Substance Abuse and Mental Health Services Administration. *Managing Chronic Pain in Adults With or in Recovery From Substance Use Disorders.* Treatment Improvement Protocol (TIP) Series 54. HHS Publication No. (SMA) 12-4671. Rockville, MD: Substance Abuse and Mental Health Services Administration, 2011.

Originating Office

Quality Improvement and Workforce Development Branch, Division of Services Improvement, Center for Substance Abuse Treatment, Substance Abuse and Mental Health Services Administration, 1 Choke Cherry Road, Rockville, MD 20857.

HHS Publication No. (SMA) 12-4671
Printed 2012

Contents

Exhibits

Consensus Panel

Chair

Margaret Kotz, D.O.
Professor, Psychiatry and Anesthesiology
Case Western Reserve University School
    of Medicine
Director, Addiction Recovery Services
Case Medical Center
University Hospitals of Cleveland
Cleveland, Ohio

Panelists

Michael Clark, M.D., M.P.H.
Associate Professor
Department of Psychiatry and Behavioral
    Sciences
Johns Hopkins University School of Medicine
Baltimore, Maryland

Peggy Compton, RN, Ph.D., FAAN
Associate Professor
UCLA School of Nursing
Los Angeles, California

Edward Covington, M.D.
Director, Neurological Center for Pain
Neurological Institute
Cleveland Clinic Foundation
Cleveland, Ohio

Carmen R. Green, M.D.
Associate Professor, Anesthesiology
Associate Professor, Health Management and
    Policy
Director, Health Disparities Research
Michigan Institute for Clinical and Health
    Research
Director, Pain Medicine Research
University of Michigan
Ann Arbor, Michigan

Joseph O. Merrill, M.D., M.P.H.
Clinical Assistant Professor of Medicine
University of Washington
Harborview Medical Center
Seattle, Washington

Steven D. Passik, Ph.D.
Associate Attending Psychologist
Department of Psychiatry and Behavioral
    Sciences
Memorial Sloan Kettering Cancer Center
Associate Professor of Psychiatry
Weill Medical College of Cornell
University Medical Center
New York, New York

Charles A. Simpson, D.C., DABCO
Vice President, Medical Director
The CHP Group
Beaverton, Oregon

What Is a TIP?

Treatment Improvement Protocols (TIPs) are developed by the Center for Substance Abuse Treatment (CSAT), part of the Substance Abuse and Mental Health Services Administration (SAMHSA) within the U.S. Department of Health and Human Services (HHS). Each TIP involves the development of topic-specific best-practice guidelines for the prevention and treatment of substance use and mental disorders. TIPs draw on the experience and knowledge of clinical, research, and administrative experts of various forms of treatment and prevention. TIPs are distributed to facilities and individuals across the country. Published TIPs can be accessed via the Internet at http://www.kap.samhsa.gov.

Although each consensus-based TIP strives to include an evidence base for the practices it recommends, SAMHSA recognizes that behavioral health is continually evolving, and research frequently lags behind the innovations pioneered in the field. A major goal of each TIP is to convey "front-line" information quickly but responsibly. If research supports a particular approach, citations are provided.

TIP Format

Most of the research that forms the evidence basis for a particular TIP is not provided in the TIP itself. Rather, those who wish to review the supporting research can access a bibliography and literature review via the Internet at http://www.kap.samhsa.gov. These online resources include links to abstracts; the online bibliography and literature review are updated every 6 months for 5 years after publication of the TIP.

TIPs focus on how-to information. Coverage of topics is limited to what the audience needs to understand and use to improve treatment outcomes. TIPs increasingly use quick-reference tools such as tables and lists in lieu of extensive text discussion, making the information more readily accessible and useful for treatment providers.

TIP Development Process

TIP topics are based on the current needs of behavioral healthcare professionals and other medical care practitioners for information and guidance. After selecting a topic, SAMHSA invites staff members from Federal agencies and national organizations to be members of a resource panel that reviews an initial draft prospectus and outline and recommends specific areas of focus as well as resources that should be considered in developing the content for the TIP. These recommendations are communicated to a consensus panel composed of experts on the topic who have been nominated by their peers. In partnership with Knowledge Application Program writers, consensus panel members participate in creating a draft document and then meet to review and discuss the draft. The information and recommendations on which they reach consensus form the foundation of the TIP. The panel Chair ensures that the guidelines mirror the results of the group's collaboration.

A diverse group of experts closely reviews the draft document. Once the changes recommended by these field reviewers have been incorporated, the TIP is prepared for publication, in print and online.

Foreword

The Treatment Improvement Protocol (TIP) series fulfills the Substance Abuse and Mental Health Services Administration's (SAMHSA's) mission to improve prevention and treatment of substance use and mental disorders by providing best practices guidance to clinicians, program administrators, and payers. TIPs are the result of careful consideration of all relevant clinical and health services research findings, demonstration experience, and implementation requirements. A panel of non-Federal clinical researchers, clinicians, program administrators, and patient advocates debates and discusses their particular area of expertise until they reach a consensus on best practices. This panel's work is then reviewed and critiqued by field reviewers.

The talent, dedication, and hard work that TIPs panelists and reviewers bring to this highly participatory process have helped bridge the gap between the promise of research and the needs of practicing clinicians and administrators to serve, in the most scientifically sound and effective ways, people in need of behavioral health services. We are grateful to all who have joined with us to contribute to advances in the behavioral health field.

Pamela S. Hyde, J.D.
Administrator
Substance Abuse and Mental Health Services Administration

Dr. H. Westley Clark, M.D.,
J.D., M.P.H., CAS, FASAM
Director
Center for Substance Abuse
Treatment
Substance Abuse and
Mental Health Services
Administration

Francis M. Harding
Director
Center for Substance Abuse
Prevention
Substance Abuse and
Mental Health Services
Administration

A. Kathryn Power, M.Ed.
Director
Center for Mental Health
Services
Substance Abuse and
Mental Health Services
Administration

1 Introduction

Chronic Pain Impact

Chronic noncancer pain (CNCP) is common in the general population as well as in people who have a substance use disorder (SUD) (Exhibit 1-1). Chronic pain is not harmless; it has physiological, social, and psychological dimensions that can seriously harm health, functioning, and well-being. As a multidimensional condition with both objective and subjective aspects, CNCP is difficult to assess and treat. Although CNCP can be managed, it usually cannot be completely eliminated. When patients with CNCP have comorbid SUD or are recovering from SUD, a complex condition becomes even more difficult to manage.

Exhibit 1-1 Statistics on Substance Use and Chronic Pain in the United States

Category	Statistic
Chronic pain patients who may have addictive disorders	32% (Chelminski et al., 2005)
People ages 20 and older who report pain that lasted more than 3 months	56% (National Center for Health Statistics, 2006)
People experiencing disabling pain in the previous year	36% (Portenoy, Ugarte, Fuller, & Haas, 2004)
People ages 65 and older who experience pain that has lasted more than 12 months	57% (National Center for Health Statistics, 2006)
Civilian, noninstitutionalized U.S. residents ages 12 and older who report nonmedical use* of pain relievers in past year	5% (Substance Abuse and Mental Health Services Administration [SAMHSA], 2007)
People ages 12 and older who report that they initiated illegal drug use with pain relievers	19% (SAMHSA, 2008)
People with opioid addiction who report chronic pain	29–60% (Peles, Schreiber, Gordon, & Adelson, 2005; Potter, Shiffman, & Weiss, 2008; Rosenblum et al., 2003; Sheu et al., 2008)

Nonmedical use is use for purposes other than that for which the medication was prescribed.

Audience

This Treatment Improvement Protocol (TIP) is for primary care providers who treat or are likely to treat adult patients with or in recovery from SUDs who present with CNCP. Given the prevalence of CNCP in the population, this audience includes virtually all primary care providers. Addiction specialists, psychiatrists, nurses, and other clinicians may find information here that will help them ensure that their patients with CNCP receive adequate pain treatment. By providing a shared basic understanding of and a common language for these two chronic conditions, this TIP facilitates cooperation and communication between healthcare professionals treating pain and those treating addiction.

Purpose

This TIP equips clinicians with practical guidance and tools for treating CNCP in adults with histories of SUDs. It does not describe how to treat SUDs or other behavioral health disorders in patients with CNCP; however, it provides readers with information about SUD assessments and referrals for further evaluation. For patients with histories of SUDs, the most controversial and possibly hazardous pain treatment in widespread use is opioid treatment. For this reason, this topic receives significant attention in Chapters 3 and 4.

Definitions

Many terms important to the treatment of CNCP in people with SUDs are used inconsistently. Clinicians should not assume that their definitions of *addiction*, *CNCP*, *physical dependence*, *recovery*, *tolerance*, or other terms are shared by others, especially by patients and their families.

It is especially important that clinicians clarify with their patients terms related to substance use. For example, patients with histories of SUDs who are no longer using substances may or may not consider themselves to be in recovery. Likewise, some mutual-help groups may not regard patients as abstinent if they are treated for SUDs with medications such as naltrexone, buprenorphine, or methadone. Many people equate physical dependence or tolerance with addiction. However, if clinicians prescribing opioids for CNCP equate these terms, they may misdiagnose their patients on opioids as having an addiction, when in fact they do not.

In 2001, the American Academy of Pain Medicine, the American Pain Society, and the American Society of Addiction Medicine formed a Liaison Committee on Pain and Addiction to standardize the use of the terms *addiction*, *physical dependence*, and *tolerance* among pain professionals. Shared understandings of these and other terms facilitate research, advance dialog among professionals in the fields of addiction and pain, and help patients make informed decisions about their treatment.

Definitions used in this TIP are presented below.

- **addiction.** A primary, chronic, neurobiologic disease, with genetic, psychosocial, and environmental factors influencing its development and manifestations. It is characterized by behaviors that include one or more of the following: impaired *control* over drug use or compulsive use, *continued use* despite harm, and *craving* (Savage et al., 2003); clinicians commonly refer to these behaviors as the "3Cs."

- **addictive substance.** The phrase *addictive substance* is controversial. The phrase draws attention to the properties of the substance; however, some experts prefer to emphasize the importance of individual variability, environment, and situational factors in addiction. Evidence suggests that animals will self-administer all drugs commonly sought by humans (with the exception of hallucinogens). Evidence also suggests that, if animals are exposed to a sufficient dose for a sufficient time, a substantial percentage will develop behaviors remarkably similar to those that suggest addiction in humans (e.g., "drug seeking" despite electrical shocks). Nonrewarding drugs (see Neurobiology of Addiction, below) do not elicit these behaviors in animals or humans. In this TIP, drugs or medications that elicit "drug seeking" behaviors are referred to as *addictive*.

- **behavioral health.** The term comprises substance use issues, mental health issues, and the prevention of both.

- **chronic noncancer pain (CNCP).** Pain that is (1) unassociated with an imminently terminal condition, and (2) unlikely to abate as a result of tissue healing, thus requiring long-term management. The term often refers to pain not caused by ongoing tissue pathology (e.g., backache, fibromyalgia). The term is problematic because it includes pain associated with sickle cell disease or recurrent pancreatitis, in which both neurological sensitization and tissue damage, at least in part, are likely. Inflammatory arthritis, connective tissue diseases, ischemia, and other conditions cause pain that persists for years yet are not, at least initially, life threatening.

- **chronic pain syndrome.** Intractable pain of 6 months or longer, with marked alteration of behavior; depression or anxiety; marked restriction in daily activities; frequent use of medication and medical services; no clear relationship to organic disorder; and a history of multiple, non-

productive tests, treatment, and surgeries (U.S. Commission on the Evaluation of Pain, 1987). This term is used casually and imprecisely to refer to pain, distress, and dysfunction that are not fully attributable to an identifiable medical condition.

- **hyperalgesia.** An abnormally intense response to a normally noxious stimulus.

- **narcotic.** Substance used to induce narcosis or stupor. *Narcotic* is not a synonym for the opioid class of medications.

- **opioid-induced hyperalgesia.** Hyperalgesia that results from the effects of opioids on the central nervous system (CNS).

- **pain.** An unpleasant sensory or emotional experience associated with actual or potential tissue damage or described in terms of such damage (International Association for the Study of Pain, 1986). Pain is subjective and may not always be corroborated by objective data.

- **physical dependence.** A state of adaptation that is manifested by a drug-class-specific withdrawal syndrome that can be produced by abrupt cessation, rapid dose reduction, decreasing the level of the drug in the blood, or administration of an antagonist (a substance that opposes the action of the drug) (Savage et al., 2003).

- **pseudoaddiction.** A controversial term coined to describe aberrant drug-related behaviors (e.g., clock watching, drug seeking), that resemble those of patients with addiction but that actually result from inadequate treatment of pain (Weissman & Haddox, 1989).

- **recovery.** A process of change through which an individual with an SUD achieves abstinence, wellness, and improved health and quality of life (Center for Substance Abuse Treatment [CSAT], 2007).

- **relapse.** A return to substance abuse after a period of abstinence.
- **substance use disorder** (SUD). A condition that includes alcohol and drug problems. SAMHSA recognizes that several terminologies (e.g., substance abuse and addiction) can be applied and respects that some individuals and communities may choose to use different terminologies (CSAT, 2007).
- **tolerance**. A state of adaptation in which exposure to a substance induces changes that result in a diminution of one or more of the substance's effects over time (Savage et al., 2003).

Pain and Addiction Basics

Studies indicate that CNCP and addiction frequently co-occur (Chelminski et al., 2005; Rosenblum et al., 2003; Savage, Kirsh, & Passik, 2008). Chronic pain and addiction have many shared neurophysiological patterns. Most chronic pain involves abnormal neural processing, which can occur at various levels of the peripheral and CNS. Similarly, the disease of addiction results when normal neural processes, primarily in the brain's memory, reward, and stress systems, are altered into dysfunctional patterns. A full understanding of each condition is still emerging, and there is much to be learned regarding neurobiologic interactions between the conditions when they co-exist.

Chronic pain and addiction are not static conditions. Both fluctuate in intensity over time and under different circumstances and require ongoing management. Treatment for one condition can support or conflict with treatment for the other; a medication that may be appropriately prescribed for a particular chronic pain condition may be inappropriate given the patient's substance use history. Other commonalities include the following:

- Both are neurobiological conditions with evidence of disordered CNS function.
- Both are mediated by genetics and environment.
- Both may have significant behavioral components.
- Both may have serious harmful consequences if untreated.
- Both often require multifaceted treatment.

Chronic pain and SUDs have similar physical, social, emotional, and economic effects on health and well-being (Green, Baker, Smith, & Sato, 2003). Patients with one or both of these conditions may report insomnia, depression, impaired functioning, and other symptoms. Effective CNCP management in patients with or in recovery from SUDs must address both conditions simultaneously (Trafton, Oliva, Horst, Minkel, & Humphreys, 2004).

Neurobiology of Pain

Both pain and responses to pain are shaped by culture, temperament, psychological state, memory, cognition, beliefs and expectations, co-occurring health conditions, gender, age, and other biopsychosocial factors. Because pain is both a sensory and an emotional experience, it is by nature subjective.

When nociceptors are excited, the stimulus is converted through transduction into action potentials that travel to the dorsal horn of the spinal cord. Signals then continue from the dorsal horn to the brain along multiple pathways in the cord: to the somatosensory cortex, where pain is evaluated; to the limbic system, where emotional reactions are mediated; to the autonomic centers that control such automatic functions as breathing, perspiration, and heart rate; and to other parts of the brain, where a behavioral response to the stimuli is determined. Nociceptive impulses are also transmitted to nearby terminals of the same nerve, where they may lead to diffuse pain and release of inflammatory substances that produce the flare and swelling that is a protective response to tissue injury (Exhibit 1-2).

Nociceptive input triggers a pain-inhibiting response. Signals traveling the ascending pathways are met by descending signals that emerge at various points along the spinal cord and brain. This antinociceptive response involves a panoply of chemicals, including endorphins, enkephalins, gamma-aminobutyric acid, norepinephrine, serotonin, oxytocin, and relaxin. Inhibitory signaling serves to attenuate nociceptive input, dampening the formation of pain sensation and providing pain relief (Brookoff, 2005).

Pain may be acute (e.g., postoperative pain), acute intermittent (e.g., migraine headache, pain caused by sickle cell disease), or chronic (persistent pain that may or may not have a known etiology). These categories are not mutually exclusive; for example, acute pain may be superimposed on chronic pain. Acute nociceptive or neuropathic pain can transform into chronic neuropathic pain in which the original sensations are extended and amplified.

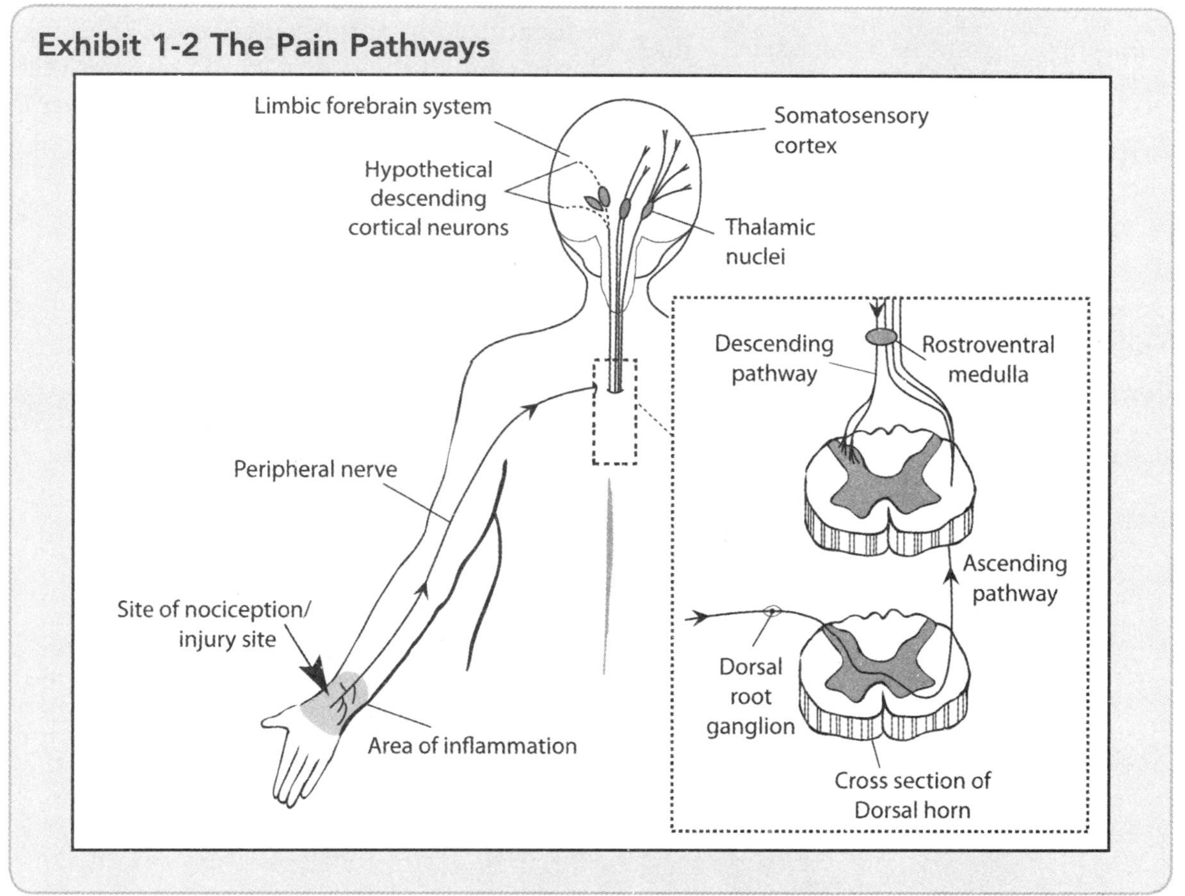

Exhibit 1-2 The Pain Pathways

Chronic Pain

Chronic pain can be nociceptive, neuropathic, or a mixture of both (Exhibit 1-3). Pains such as migraine and fibromyalgia, in which there is no noxious stimulus and no apparent neurological lesion, are attributed to dysfunction of a structurally intact CNS.

Chronic pain often results from a process of neural sensitization following injury or illness in which thresholds are lowered, responses are amplified (hyperalgesia), normally non-noxious stimulation becomes painful (allodynia), and spontaneous neural discharges occur. Increased signaling disconnected from nociceptive input can become autonomous, self-sustaining, and progressive, leading to the continuous perception of pain even in the absence of ongoing tissue damage. Thus, chronic pain is not equivalent to prolonged acute pain and for clinical purposes is best considered a distinct disorder (Brookoff, 2005).

The etiology of the abnormal processing in chronic pain is not fully understood. However, there are two main, nonexclusive causes. First, tissue damage can trigger the release of chemicals that sensitize the nerve fibers and alter gene expression, causing changes in signaling through many different mechanisms. Some of these changes enable non-pain-conducting fibers to trigger pain in the CNS. Second, pain can result from injured nerve fibers that regenerate in a neuroma, which generates pain signals with little or no stimulation.

When injury occurs to key pain-processing sectors of the CNS (e.g., the dorsal horn, thalamus), neural signals that pass through them may be interpreted as pain. Injury may also lead to degeneration of pain inhibitory cells. Modulation of nociceptive stimuli and inhibitory responses can occur at one or more locations in the CNS: the peripheral nerves, spinal cord neurons and tracts, thalamus, and cortex (Compton & Gebhart, 2003). Accurate identification of the source of the chronic pain, and of the neurological processes that modulate it, can lead to rational therapeutic approaches that target the source of aberrant signaling on the CNS pathway.

Exhibit 1-3 Pain Types

Type	Description
Nociceptive Pain	Pain that results from suprathreshold stimulation of nociceptors, which are neural receptors specialized for the detection of potentially harmful situations. This is an adaptive function of the nervous system. Nociceptors can be excited by mechanical, thermal, or chemical stimulation. The immediate physical response is reflexive and protective, causing a person to pull a hand away from a hot surface, for example. Nociceptive pain persists while the injurious agent remains or until healing occurs. Prolonged nociceptive input can cause central hypersensitization and the experience of spontaneous or amplified pain.
Neuropathic Pain	Pain that results from lesion or dysfunction of the sensory nervous system. A compressed, injured, or severed nerve can trigger neuropathic pain, as can disorders that affect the neural axis (e.g., metabolic diseases, infections, autoimmune disorders, vascular diseases, neoplasia [Campbell & Meyer, 2006]).
Mixed Nociceptive/ Neuropathic Pain	A combination of the two types of pain. For example, patients with degenerative disc disease may suffer from mechanical (nociceptive) back pain and radicular (neuropathic) pain.

Pain's Effect on Health

Persistent pain can have significant adverse effects on health. When pain stimuli continuously trigger the stress response, the acute signs of sympathetic activation (e.g., rapid heart rate, sweating) may cease or appear intermittently, yet the body continues to be stressed. This situation contributes to a sense of exhaustion.

Continued pain can trigger emotional responses, including sleeplessness, anxiety, and depressive symptoms, which in turn produce more pain. Such feedback cycles may continue to cause pain after the physiological causes have been addressed. Several studies show that the outcome of pain treatment is worse in the presence of depression, or when depression does not respond to treatment, and that the future course of pain syndromes can be, in part, predicted by emotional status. The physiological and psychological sequelae of CNCP can be exacerbated by such factors as inactivity and overuse of sedating drugs. Physical inactivity and a lack of engagement with life may also lead to increased levels of anxiety, depression, and an increased risk for suicidal ideation; these increases may lead a person to use substances in an attempt to treat these sequelae of CNCP and the losses that occur due to its presence.

Neurobiology of Addiction

A person may use substances initially for several reasons, such as to experience the euphoric effects, to relieve stress, to overcome anxiety or depression (or both), or to blunt the pain (National Institute on Drug Abuse [NIDA], 2007). With repeated exposure, however, substance use in some people can become uncontrollable. The defining characteristics of the disease of addiction have been summarized as the "3Cs" (see the definition of *addiction*). Changes to the brain occur in a process that is mediated by both genetic and environmental factors, which result in an overvaluation of the substance, a devaluation of other things, and impaired control of substance-related behavior. Evidence indicates that addiction is a chronic disease.

The primary rewarding effects of addictive substances occur in the cortico-mesolimbic dopamine systems, where several structures link to control the basic emotions and connect them to memories, which drive behavior. These systems produce sensations of pleasure in response to actions that support survival (e.g., eating, sex) and sensations of fear in response to potential dangers. In a cascading effect, these sensations trigger the endocrine and autonomic nervous systems, stimulating bodily responses. The prefrontal cortex also plays a role in the formation of addictions, modifying pleasure and pain signals based on other considerations. Thus, the brain's reward and stress systems reinforce life-sustaining behaviors.

Reward Response

Feelings of reward emerge from the core of the limbic system after neurons in the ventral tegmental area (VTA) release the neurotransmitter dopamine into the nucleus accumbens (NAc). Neural activity within this VTA–NAc circuit is necessary to experience reward, but other areas within the broader brain reward circuit also exert a strong influence. For example, the hippocampus contributes information from the past that may be relevant to the current experience. The amygdala adds critical information about the emotional valence of the stimulus that activated the reward circuit, thus contributing to the overall motivational power of the experience. In addition, parts of the prefrontal cortex (i.e., anterior cingulate and orbitofrontal cortices) help integrate all the information, a vital function that allows the individual to decide whether to initiate or suppress a particular behavior in response to the stimulus.

Most addictive substances increase the levels of dopamine in limbic targets well beyond what occurs in naturally rewarding situations (e.g., sex, food). Some drugs (e.g., marijuana, heroin) produce dopaminergic effects indirectly. Amphetamines cause the release of dopamine, and cocaine prevents its reuptake; both effects result in amplified messaging that eventually disrupts normal neuronal signaling.

It appears that the brain adjusts to excess dopamine levels by producing less dopamine and by reducing the number of receptors that respond to it in the receiving (postsynaptic) neuron. As a result, the pleasurable effects of a drug become diminished with continued use. The pleasurable effects of normal activities also are blunted, creating a state called *anhedonia* (an inability to experience pleasure).

It is commonly believed that continued substance use is driven by the need to prevent symptoms of withdrawal; however, this idea is misleading. Withdrawal, as commonly conceptualized, involves rebound symptoms resulting from the drug's absence. In the case of opioids, these symptoms include, but are not limited to, anxiety, sweating, tachycardia, diarrhea, piloerection, and chills. Although unpleasant, these symptoms are typically absent after a relatively brief period of detoxification or weaning and do not explain phenomena such as addiction relapse and prolonged craving. Even though detoxification is quick and technically easy, the prevention of relapse is extremely difficult and, in fact, the majority of those who attain abstinence experience at least one relapse (Dennis, Foss, & Scott, 2007). These more difficult problems are thought to result from the prolonged impairment of hedonic tone and conditioned responses that lead to intense craving.

Stress Response

The dysregulation of the brain's reward system that occurs through substance use is paralleled by similar dysregulating effects in the stress system. Use of an addictive substance increases the flow of neurochemicals (e.g., corticotropin-releasing factor, norepinephrine, dynorphin). These chemicals can produce a negative emotional state that manifests as chronic irritability, emotional pain, lethargy, disinterest in natural rewards, and other dysphoric conditions. The stress response becomes more sensitive with repeated withdrawal and can persist into abstinence (Koob, 2009).

An individual may seek to avert the stress response by again using the substance. This negative reinforcement combines with the positive reinforcement of the substance's euphoric effects in an operant process that creates a compulsion for substance use. Thus, addiction is reflected in compulsive use combined with loss of control mediated by memory (cue-induced triggers for reuse), substance-induced reductions in executive functioning that hamper rational decisionmaking, and habit formation (Koob, 2009).

Risk Factors for Addiction

People who use substances with addictive potential may develop tolerance to some of their effects and develop some degree of physical dependence. However, only a minority develops the disease of addiction. Important risk factors for addiction include genetics, psychological factors, and environmental factors.

Genetics plays a substantial role in risk factors for addiction: NIDA (2007) estimates that between 40 and 60 percent of a person's vulnerability to addiction may be genetic. The disease of addiction may be more heritable than type 2 diabetes, hypertension, and breast cancer (Nestler, 2005). Genes also underlie human variability in drug metabolism, susceptibility to psychiatric disorders that commonly co-occur with addiction, and response to environmental risk factors (e.g., drug availability, peer group pressure [Vaillant, 2003]).

Mental illness is another major risk factor for addiction (Volkow & Li, 2009), and the two conditions have high comorbidity (NIDA, 2009). One condition can follow the other (NIDA, 2007). For example, a person may attempt to relieve depression or anxiety with substances, and this behavior may lead to addiction. Conversely, chronic substance use may lead to mental disorders, such as psychosis, or make existing mental illness worse (NIDA, 2007). Environmental influences on addiction include, but are not limited to, poverty, poor parental support, living in a community with high drug availability, and using substances at an early age (NIDA, 2007, 2009; Volkow & Li, 2009).

Cross-Addiction

Addiction to one substance can be linked with addiction to other substances in a pattern termed *cross-addiction*. An individual who voluntarily or involuntarily decreases use of one substance may increase use of another substance with similar effects on the brain (e.g., the person with an alcohol use disorder may use barbiturates for the sedative effects). The term cross-addiction is also used to describe simultaneous addiction (e.g., co-occurring addictions to nicotine, alcohol, and marijuana).

Cross-addiction is not official diagnostic nomenclature; rather, it refers to the observation that a person with an addiction to one substance may develop addiction to a subsequent substance, especially if the original drug of choice becomes inaccessible or is

relinquished for other reasons. For example, a study of patients hospitalized for controlled-release oxycodone addiction found that the majority (77 percent) had previously had a non-opioid SUD (Potter, Hennessey, Borrow, Greenfield, & Weiss, 2004).

Individuals with chronic pain and histories of SUDs may be at increased risk of cross-addiction to any medication that acts on the brain as a reinforcing agent (Edlund, Sullivan, Steffick, Harris, & Wells, 2007). Because of cross-addiction, persons who abuse marijuana may be at increased risk for opioid addiction. People with alcohol use disorders have been found to be more than 18 times as likely to report nonmedical use of prescription medications as people who do not drink (McCabe, Cranford, & Boyd, 2006).

The Cycle of Chronic Pain and Addiction

Although multiple factors influence the course of addiction, CNCP provides both positive and negative reinforcement of substance use. Positive reinforcement occurs when a behavior is followed by a consequence that is desirable— a donkey's walking may be rewarded by a carrot. Negative reinforcement occurs when a behavior is followed by the elimination of a negative consequence—a donkey's walking may eliminate the blows from a stick. For example, euphoria is a positive reinforcer for taking heroin, and pain reduction is a negative reinforcer for taking heroin. Prescribed opioids, benzodiazepines, or other medications may dramatically relieve pain or distress (e.g., depression, anxiety). Unprescribed substances may be used for similar reasons; for example, alcohol may promote relaxation or sleep. Such relief is a strong reinforcement for repeated consumption of the substance.

Unfortunately, analgesic and anxiolytic efficacy may diminish over the course of weeks, months, or years as tolerance develops. This loss of efficacy often elicits dose escalation to recapture efficacy. This escalation is rewarded, as the increased dose is initially more effective than the lower dose.

If the drug produces physical dependence, the person may have not only increased pain when the substance is absent, but also withdrawal symptoms (e.g., anxiety, nausea, cramps, insomnia). Withdrawal symptoms may lead to an increase in symptoms of depression and an increase in the potential risk for suicide. All these symptoms are relieved by ingesting more of the drug that caused the dependence. A similar situation may occur if the drug is one that elicits rebound symptoms. For example, ergot relieves migraine, but excessive use leads to rebound headaches that are more persistent and treatment resistant than were the original headaches.

An illusion of benefit produced by reinforcing drugs can create a paradoxical situation in which long-term use of the substance creates the very symptoms the person hopes to alleviate. People commonly drink to relax or "cheer up," yet chronic alcohol abuse leads to depression and anxiety.

In some people, a cycle develops in which pain or distress elicits severe preoccupation with the substance that previously provided relief. This cycle—seeking pain relief, experiencing relief, and then having pain recur—can be very difficult to break, even in the person without an addiction, and the development of addiction markedly exacerbates the difficulty. The propensity to develop this cycle is influenced by genetic and environmental factors; some people will experience greater degrees of analgesia than others, and some will have more severe or prolonged abstinence symptoms. Genetic variability in susceptibility to these experiences may explain some cases of iatrogenic addiction.

Summary of TIP

The management of CNCP in patients with a comorbid SUD is challenging for both patients and clinicians; however, it can be done successfully. This TIP advises clinicians to conduct a careful assessment; develop a treatment plan that addresses pain, functional impairment, and psychological symptoms; and closely monitor patients for relapse. Even the best treatment is unlikely to completely eliminate chronic pain, and efforts to achieve total pain relief can be self-defeating. Patients may benefit when clinicians team with other professionals (e.g., psychologists, addiction counselors, pharmacists, holistic care providers). Patients must also assume a significant amount of responsibility for optimal management of their pain. Educating patients, family members, and caregivers in this process, and helping patients improve their quality of life, can be gratifying for everyone involved.

Key Points

- CNCP and the disease of addiction involve neurophysiological processes.
- Both genetic and environmental factors contribute to and influence the development and course of CNCP and addiction.
- Clinicians must understand CNCP, addiction, and other behavioral health issues to best serve the chronic pain patient with or in recovery from an SUD.
- Despite the complexities of CNCP and SUDs, patients with these co-occurring, chronic conditions can be treated effectively.

2 Patient Assessment

Elements of Assessment

Researchers and clinicians agree that, because chronic noncancer pain (CNCP) is a multifaceted condition, assessment must include more than measures of pain intensity (Brunton, 2004; Haefeli & Elfering, 2006; Karoly, Ruehlman, Aiken, Todd, & Newton, 2006; Sullivan & Ferrell, 2005). Some elements are essential to assess; others, ideal. In many cases, even after a thorough assessment, the clinician may not detect the nociceptive source of a patient's chronic pain.

Collateral information is an important part of the assessment. Clinicians need to communicate with families, pharmacists, and other clinicians after the patient has given full consent for these discussions. If the patient declines to give consent, prolonged treatment with controlled substances may be contraindicated. Furthermore, a clinician who prescribes controlled substances to a patient who refuses to permit access to outside information could be considered to be ignoring evidence of addiction or substance misuse and, therefore, to be trafficking. Collateral information also helps protect the patient from misusing medications. Exhibit 2-1 presents elements of a comprehensive assessment.

Assessment Tools

Standardized instruments provide ways to assess and track patient pain levels, function, substance use, and other factors important to managing CNCP. Standardized tools provide supplemental information for treatment planning and assessment of risk and outcomes. If used well, tools can reduce clinician bias during patient assessment.

The sensitivity and specificity of screening instruments vary, and all can yield false-positive or false-negative results. In addition, no single instrument has been shown to be appropriate for use with all patient populations (Bird, 2003; Brunton, 2004). Because of their limitations, standardized tools should not be the absolute determinants of treatments offered or withheld.

Exhibit 2-1 Elements of a Comprehensive Patient Assessment

Element	Assessment Factor
Pain and Coping	• Location, character (e.g., shooting or stinging, continuous or intermittent) • Pain types (i.e., nociceptive, neuropathic, mixed) • Lowest and highest extent of pain in a typical day, on a 0-to-10 scale • Usual pain in a typical day, on a 0-to-10 scale • When and how the pain started • Exacerbating factors (e.g., exertion/activity, food consumption, elimination, stress, medical issues) • Palliating factors (e.g., heat, cold, stretching, rest, medications, complementary and alternative treatments) • Prior evaluations to determine the source of pain • Response to previous pain treatments, including complementary and alternative treatments and interventional treatments • Goals and expectations for pain relief
Collateral Information	It is crucial to obtain such information as: • Findings of other clinicians, prior and current • Family concerns, beliefs, and observations • Pharmacist concerns, where relevant • Data from State electronic prescription monitoring programs, if available • Medical records, including psychiatric and substance use disorders (SUDs) treatment records
Function	Effect of pain on: • Activities of daily living/ability to care for oneself • Sleep • Mood • Work/household responsibilities • Sex • Socialization and support systems • Recreation • Goals and expectations for restored function
Contingencies	• Family support of wellness versus illness behavior • Vocational incentives and disincentives • Financial incentives and disincentives • Insurance/legal incentives and disincentives • Environmental and social resources for wellness

Exhibit 2-1 Elements of a Comprehensive Patient Assessment (continued)

Element	Assessment Factor
Substance Use History and Risk for Addiction	• Current use of substances, including tobacco, alcohol, over-the-counter medications, prescription medications, and illicit drugs (confirmed by toxicology) • Focus on opioids to the exclusion of other treatments • Adverse consequences of use (e.g., functional impairment; legal, social, financial, family, work, medical problems) • Age at first use • Treatment history, including attendance at mutual-help groups • Periods of abstinence • Strength of recovery support network (e.g., sponsor, sober support network, mutual-help meetings) • Family history of SUD • History of physical, sexual, or emotional abuse or trauma
Co-Occurring Conditions and Disorders	• Psychological conditions (e.g., depression, anxiety, post-traumatic stress disorder [PTSD], somatoform disorders) • Medical conditions (e.g., hepatic, renal, cardiovascular, metabolic) • Cognitive impairments (e.g., dementia, delirium, intoxication, traumatic brain injury)
Physical Exam	• Relevant associated signs of pain disorder • Signs of substance abuse (e.g., track marks, hepatomegaly, residua of skin infections, nasal and oropharyngeal pathology)
Mental Status	• Medication focused • Somatic preoccupation • Mood • Suicidal ideation and behavior • Cognition (e.g., attentional capacity, memory)

When using standardized tools, clinicians should (Bird, 2003):

- Understand the strengths and weaknesses of each tool.
- Select a tool appropriate for the patient, considering memory problems, cognitive impairments, eyesight, literacy level, cultural background, gender, ethnicity, and other factors.
- Teach patients how to use self-administered tools, even "self-explanatory" tools; otherwise, the information they provide may be invalid.

Instruments are available to assist with assessment of pain and functioning, SUDs, psychiatric comorbidities, coping skills, and potential problems with opioid use.

Assessing Pain and Function

The assessment of CNCP should include documentation of the following:

- Pain onset, quality, and severity; mitigating and exacerbating factors; and the results of investigations into etiology
- Pain-related functional impairment

- Emotional changes (e.g., anxiety, depression, anger) and sleep disturbances
- Cognitive changes (e.g., attentional capacity, memory)
- Family response to pain (i.e., supportive, enabling, rejecting)
- Environmental consequences (e.g., disability income, loss of desired activities, absence from desirable or feared work)
- Physical examination
- Partial mental status examination (e.g., affect [how pain is experienced], somatic preoccupation, cognition, moans, gasps, lying down during the interview)

Several factors may complicate an assessment of pain levels in any pain patient:

- Some patients may report not only their level of pain intensity, but their suffering, which may be greater than their pain intensity.
- Clinicians tend to believe that a patient's pain level is actually lower than the patient reports, except when the patient reports low pain (Sloman, Rosen, Rom, & Shir, 2005; Stalnikowicz, Mahamid, Kaspi, & Brezis, 2005).
- Clinicians are especially likely to underestimate—and, therefore, to undertreat—pain and disability in women, the elderly, minorities, people of low economic status, and people with SUDs (Green, Baker, Smith, & Sato, 2003; Rupp & Delaney, 2004).

An assessment of pain and function in patients with SUD histories may be further complicated by the following factors:

- Some patients with histories of SUDs may overreport their pain experience if they are afraid that they will be undermedicated or that their symptoms will not be taken seriously.

- Others may underreport their pain experience if they are afraid they will be prescribed medications that will cause them to relapse.
- Some patients may exaggerate pain and disability levels to get opioids for reasons other than pain control.

The level of functional impairment in patients with CNCP is markedly modified by environmental contingencies (e.g., the incentives and disincentives for healthful versus so-called "sick role" behaviors). For instance, evidence shows that pain-related behaviors increase in the presence of a solicitous spouse, meaning one who is attentive to and reinforcing of such behaviors (Pence, Thorn, Jensen, & Romano, 2008). It is also demonstrated that work-related functional impairment varies with the strengths of reinforcement contingencies for function versus absenteeism. The workers' compensation system may provide a special example of this. Studies typically find that patients receiving income from this source respond less well to rehabilitation efforts than do those not receiving disability income from this or other sources. The explanation is thought to reside in such factors as the need to "prove" one is ill to obtain tests and specialty consultation and the fear of loss of income if one is witnessed engaging in normal activities. The relative magnitude of rewards and punishments for function may thus play a determining role in disability. A thorough assessment of a patient with CNCP, therefore, requires a review of the overall consequences of resuming healthy function.

When assessing pain and function in patients with histories of SUDs, clinicians should keep in mind the following:

- Individuals with similar complaints (e.g., low back pain) usually describe and rate their pain differently.
- Functional impairments affect patients differently.

- Pain scores do not reflect tissue pathology, disability, or treatment response.
- Pain reduction is insufficient to judge treatment success, which also requires optimization of function and normalization of mood.

Exhibits 2-2 and 2-3 list the strengths and weaknesses of common one-dimensional and multidimensional pain tools, respectively. Exhibit 2-4 presents tools for assessing the extent to which pain interferes with usual functions and activities. Information on how to obtain the tools is located in Appendix B.

Studies show that patients who have chronic pain may develop cognitive impairments (e.g., changes in attentional capacity, memory, processing speed) that appear to be independent of other variables (e.g., age, educational level, pain intensity, pain relief) (Dick & Rashtiq, 2007; Hart, Martelli, & Zasler, 2000; Hart, Wade, & Martelli, 2003). Therefore, clinicians need to be alert to the possibility of these changes and include an evaluation of mental status as part of the patient's ongoing assessment (e.g., the *Mini–Mental State Examination,* [Folstein & Folstein, 2010]) or refer the patient to a neurologist as necessary.

Exhibit 2-2 Tools To Assess Pain Level

Tool	Strength	Weakness
Faces Pain Scale	• Easy to use • Usable with people who have mild to moderate cognitive impairment • Translates across cultures and languages	• Visual impairment may affect accuracy or completion • May measure pain affect, not only pain intensity
Numeric Rating Scale (NRS)	• Easy to use if patient can translate pain into numbers • Easy to administer and score • Can measure small changes in pain intensity • Oral or written administration • Sensitive to changes in chronic pain • Translates across cultures and languages	• Difficult to administer to patients with cognitive impairments because of difficulty translating pain into numbers
Verbal Rating Scale/Graphic Rating Scale	• Easy to use • Oral or written administration • High completion rate with patients with cognitive impairments • Sensitive to change and validated for use with chronic pain • Correlates strongly with other tools	• Not as sensitive as NRS or Visual Analog Scale

---

Exhibit 2-2 Tools To Assess Pain Level (continued)

Tool	Strength	Weakness
Visual Analog Scale (VAS)	• Easy to use, but must be presented carefully • Precise • Sensitive to ethnic differences • Easily translated across cultures and languages • Some evidence that a horizontal line may be better than a vertical ("thermometer") orientation	• Visual impairment may affect accuracy • Can be time consuming to score, unless mechanical or computerized VAS tools are used • Low completion rate in patients with cognitive impairments • Difficult to administer to patients with cognitive impairments • Cannot be administered by phone or email • Subject to measurement error

Bird, 2003; Brunton, 2004.

Exhibit 2-3 Tools To Assess Several Dimensions of Pain

Tool	Strength	Weakness
Brief Pain Inventory	• Short form better for clinical practice • Fairly easy to use • Useful in different cultures • Translated into and validated in several languages	• Not easily used with patients with cognitive impairments
McGill Pain Questionnaire	• Short form easier to administer • Extensively studied	• Measures pain affect • Not appropriate for patients with cognitive impairments • Translation complicated • Meaning of pain descriptors may vary across racial and ethnic groups

Department of Veterans Affairs & Department of Defense, 2003.

Assessing Substance Use and Addiction

As with assessing pain and function, assessing patient self-reports of substance use, whether via interviews or structured self-report questionnaires, should be corroborated by other sources of information (e.g., medical records, interviews with family, urine toxicology, information from State prescription monitoring programs) (Katz & Fanciullo, 2002).

Exhibit 2-4 Tools To Assess Pain Interference With Life Activities and Functional Capacities

Tool	Purpose
Katz Basic Activities of Daily Living Scale	Rates independence by assessing six areas of daily activities
Pain Disability Index	Measures chronic pain and chronic pain interference in daily life
Roland-Morris Disability Questionnaire	Measures perceived disability from low back pain
WOMAC Index	Assesses pain, stiffness, and physical function in patients with osteoarthritis

---

When initiating a conversation about alcohol and drug use, clinicians should:

- Approach the topic matter-of-factly, handling it as part of the overall medical history.

- Incorporate questions about drug and alcohol use into a general behavioral health inventory including discussion of other lifestyle behaviors (e.g., diet, exercise).

- Ask about nicotine and caffeine use; questions about use of these substances provide opportunities to move to assessment of other substances, beginning with alcohol, the most commonly abused substance.

- Assure patients that honest answers to questions of substance use are necessary to developing a treatment plan and that their responses will remain confidential.

A good prescreening question is, "When did you last have a drink of beer, wine, or liquor?" If the patient reports drinking within the past year, the clinician should ask questions to determine:

- Frequency ("How many days per week do you typically drink alcohol?")

- Quantity ("How much alcohol do you drink on a typical drinking occasion?")

- Evidence of binge drinking (for men: "On any day in the past year, have you ever had five or more drinks?"; for women: "On any day in the past year, have you ever had four or more drinks?")

The clinician should ask the patient to define what the patient means by "a drink" (e.g., an 8-ounce glass, half a glass). The National Institute on Alcohol Abuse and Alcoholism (NIAAA) defines one drink as one 12-ounce bottle of beer or wine cooler, one 5-ounce glass of wine, or 1.5 ounces of 80-proof distilled spirits. According to NIAAA (2005), if the male patient drinks more than 4 standard drinks in a day (or more than 14 drinks per week), or more than 3 drinks in a day (or more than 7 drinks per week) for the female patient, the person is at increased risk for developing alcohol-related problems.

Whether or not the patient reports drinking, the clinician should probe for the use of licit and illicit drugs, starting with the most commonly used illicit drug in the United States: marijuana. Questions can continue to address other major classes of drugs with abuse potential (e.g., depressants, stimulants, opioids), with particular attention to use related to controlling pain or the patient's anxiety and fear of pain (Passik & Kirsch, 2004). Exhibit 2-5 summarizes the substances that patients should be asked about using.

NIDA provides a Web-based tool that helps clinicians screen for tobacco, alcohol, and illicit and nonmedical prescription drug use, and suggests levels of intervention. The tool is at http://ww1.drugabuse.gov/nmassist.

Screening for Substance Use Disorders

Although the amount of substance used is significant, it is more important to evaluate the consequences of the drug and alcohol use on life domains, such as family, work or school, and involvement with the criminal justice system (e.g., arrests for driving under the influence). When drug or alcohol use interferes with normal function, addiction is likely. Furthermore, addiction is characterized by impaired ability to control use of the substance. Asking whether the patient has ever attempted to decrease the amount consumed is an approach to determining his or her ability to modulate use. In the case of prescription medication, a patient's loss of control may manifest as the inability to ration pills until the next prescription, so the patient's partner may oversee the dispensing of the medications.

The *Diagnostic and Statistical Manual of Mental Disorders,* Fourth Edition, Text Revision (DSM-IV-TR; American Psychiatric Association [APA], 2000) provides criteria for determining substance dependence that enable the clinician to distinguish between patients with at-risk substance use and those whose use is consistent with an SUD (Exhibit 2-6). It is important to remember that, essentially, all patients taking prescribed opioids or sedatives on a long-term basis will have a degree of tolerance and withdrawal and that these criteria are not indicative of addiction absent the "maladaptive pattern of substance use."

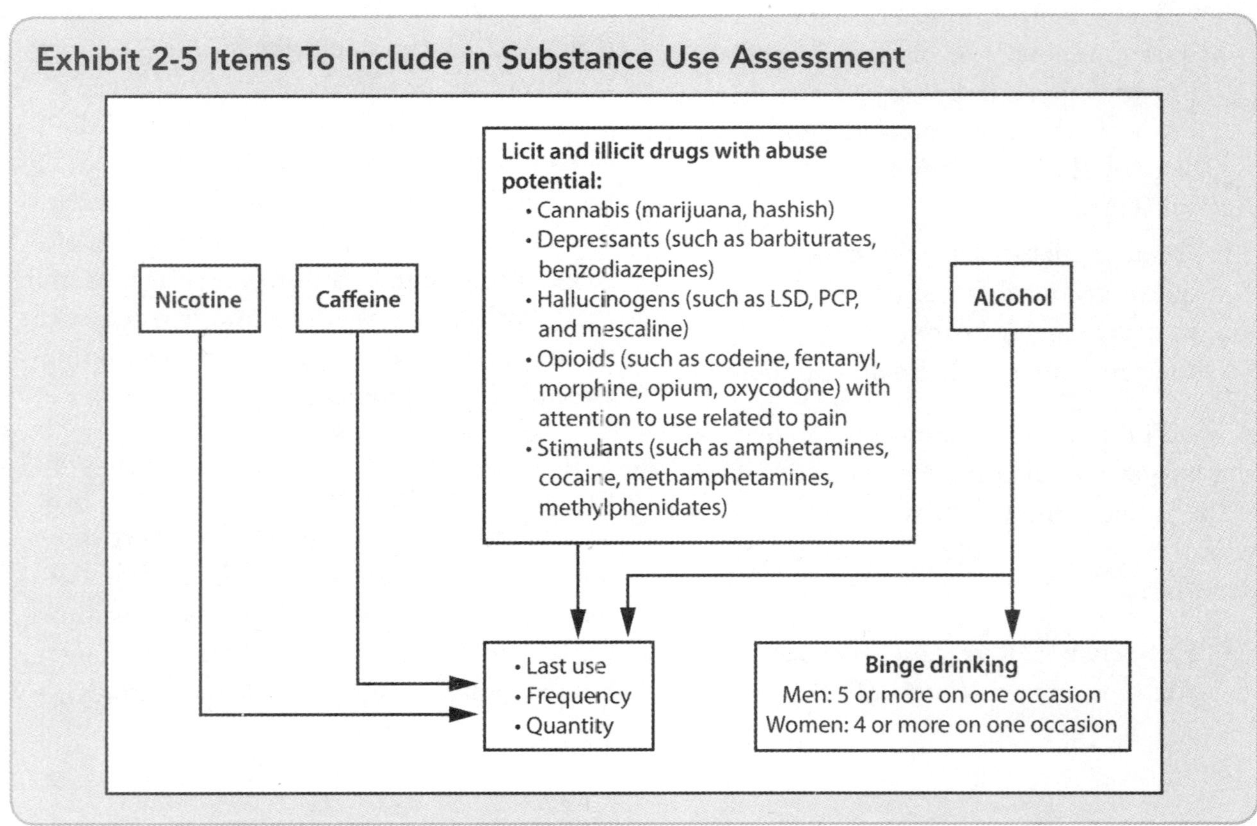

Exhibit 2-5 Items To Include in Substance Use Assessment

Exhibit 2-6 DSM-IV-TR Criteria for Substance Abuse and Substance Dependence

Category	Criteria
Substance Abuse	A maladaptive pattern of substance use leading to clinically significant impairment or distress, as manifested by one (or more) of the following, occurring within a 12-month period: • Recurrent substance use resulting in a failure to fulfill major role obligations at work, school, or home (e.g., repeated absences or poor work performance related to substance use; substance-related absences, suspensions, or expulsions from school; neglect of children or household) • Recurrent substance use in situations in which it is physically hazardous (e.g., driving an automobile or operating machinery when impaired by substance use) • Recurrent substance-related legal problems (e.g., arrests for substance-related disorderly conduct) • Continued substance use despite having persistent or recurrent social or interpersonal problems caused or exacerbated by the effects of the substance (e.g., arguments with spouse about consequences of intoxication, physical fights)
Substance Dependence	A maladaptive pattern of substance use, leading to clinically significant impairment or distress, as manifested by three (or more) of the following, occurring any time in a 12-month period: • Tolerance, as defined by either of the following: (a) a need for markedly increased amounts of the substance to achieve intoxication or desired effect, or (b) markedly diminished effect with continued use of the same amount of the substance • Withdrawal, as manifested by either of the following: (a) the characteristic withdrawal syndrome for the substance, or (b) the same (or closely related) substance is taken to relieve or avoid withdrawal symptoms • The substance is often taken in larger amounts or over a longer period than intended • There is a persistent desire or unsuccessful efforts to cut down or control substance use • A great deal of time is spent in activities necessary to obtain the substance, use the substance, or recover from its effects • Important social, occupational, or recreational activities are given up or reduced because of substance use • The substance use is continued despite knowledge of having a persistent physical or psychological problem that is likely to have been caused or exacerbated by the substance (e.g., current cocaine use despite recognition of cocaine-induced depression, continued drinking despite recognition that an ulcer was made worse by alcohol consumption)

Reprinted with permission from the *Diagnostic and Statistical Manual of Mental Disorders,* Fourth Edition, Text Revision (Copyright 2000). American Psychiatric Association.

Although a patient's former drug of choice is the one that is most likely to lead to cravings and relapse (Daley, Marlatt, & Spotts, 2003; Gardner, 2000), clinical experience suggests that a person with a history of an SUD involving any drug is susceptible to developing a cross-addiction with opioids (Covington, 2008; Savage, 2002).

Clinicians should try to determine patients' recovery status, which is crucial in developing a treatment plan (Exhibit 2-7). Many patients will be forthcoming about past or recent substance abuse during a comprehensive assessment. Some patients who have an SUD lack a full appreciation of the effects of substances, prescribed or otherwise, on their function; however, family members can usually provide this information.

Several standardized tools for SUD screening are listed in Exhibit 2-8. Information on how to obtain the tools is in Appendix B. Most tools are short, can be self-administered, and can be integrated into the health-screening forms the patient completes prior to seeing the clinician. Although no tool is a substitute for a good clinical interview, screening is essential to case finding and a useful complement to the patient interview, the physical exam, and ongoing observation (Fishman, 2007).

Exhibit 2-7 Steps Following Substance Abuse Assessment

If	Then
Abuse is remote and patient is in long-term recovery	Verify and support recovery efforts
Patient is on buprenorphine or methadone maintenance therapy (MMT)	Verify and continue buprenorphine or MMT
Abuse appears active	Refer patient to substance abuse specialist for further evaluation

Adapted from Passik & Kirsh, 2004.

Exhibit 2-8 Tools To Screen for Substance Use Disorders

Tool	Format	Administration/ Scoring Time	Training Required
Alcohol, Smoking, and Substance Involvement Screening Test	1 item for lifetime use, 6 items for each of 10 substances used, and 1 item on injection use	Depends on number of substances used	Yes
Alcohol Use Disorders Identification Test (AUDIT)	10-item screening questionnaire	2 minutes to administer/ 1 minute to score	Yes
AUDIT-C	3-item screening questionnaire	Less than 1 minute to administer and score	Yes
CAGE Adapted To Include Drugs	4 yes/no questions	Less than 1 minute/ not scored	No
Drug Abuse Screening Test	20 yes/no questions about current and past use	1–2 minutes to administer/ not scored	No
Michigan Alcoholism Screening Test (MAST) (MAST-G for older adults)	24 yes/no questions	10 minutes to administer/ 5 minutes to score	No

The Substance Abuse and Mental Health Services Administration's Screening, Brief Intervention, and Referral to Treatment (SBIRT) initiative may be helpful in the primary care context (Exhibit 2-9). More information can be obtained from the Center for Substance Abuse Treatment (CSAT, 1999a). Research findings on SBIRT are available from National Association of State Alcohol and Drug Abuse Directors (2006).

Referring for Further Assessment

If the clinical interview, collateral interview, medical records, and screening suggest an unacknowledged SUD in a patient seeking treatment for CNCP, the clinician should refer the patient to an SUD specialist, if possible. Ideally, clinicians should develop a strong referral network of substance abuse treatment clinicians who can collaborate in the care of these high-risk patients, but specialists may not always be available or accessible. Referral for an SUD does not obviate the need for pain treatment because addiction treatment facilities rarely have the resources or expertise to treat pain.

Patients may react negatively to a referral to an SUD specialist. To avoid surprising the patient and putting the specialist in an awkward situation, the clinician should clearly explain the purpose of the referral. When referring the patient, clinicians should:

- Present the referral to the SUD specialist as they would a referral to any specialist, using a matter-of-fact and unapologetic tone.

- Explain to the patient the importance of assessing factors that may be contributing to chronic pain, including substance use, and the problems SUDs or substance use may present for optimal treatment of chronic pain.

- Avoid getting distracted by the patient's explanation of his or her substance use.

- Assure the patient that the referral does not mean transfer of care. The patient needs to know that care will be coordinated among all professionals involved, if indicated, and that discussions of short- and long-term treatment will involve everyone, including the patient.

Exhibit 2-9 Elements of Screening, Brief Intervention, and Referral to Treatment

Category	Description
Screening	Identifies individuals with problems related to substance use. Screening can be through interview and self-report.
Brief Intervention	Follows a screening result indicating a moderate risk. A successful brief intervention encompasses support of the patient's ability to make behavioral change.
Brief Treatment	Follows a screening result of moderate to high risk. Brief treatment includes assessment, education, solving problems, introducing coping mechanisms, and building a supportive social environment.
Referral to Treatment	Follows a screening result indicating severe abuse or dependence. This process facilitates access to care for individuals requiring more extensive treatment than SBIRT provides and ensures access to the appropriate level of care for all who are screened.

- Help the patient make the appointment or make the appointment for the patient.

The clinician–patient relationship is especially critical for patients who have comorbid pain and an SUD. They may anticipate that clinicians will criticize their substance use and discount their pain, and they may misinterpret a concern about an SUD as a lack of concern for their pain. They may blame themselves for having developed an SUD and expect the clinician to do the same. Therefore, the clinician must maintain an attitude of respect and concern. The clinician should assure the patient that both pain and the SUD are uninvited chronic illnesses and that both need to be treated concurrently.

Federal regulations hold clinicians to a high standard of confidentiality regarding patient drug and alcohol treatment information (Exhibit 2-10). Appendix C provides elements of a written consent and a sample consent form from 42 Code of Federal Regulations (CFR).

Psychiatric Comorbidities

Both CNCP and SUDs are associated with high rates of psychiatric comorbidities, such as anxiety, depression, PTSD, and somatoform disorders (Chelminski et al., 2005; Dersh, Polatin, & Gatchel, 2002; Lebovits, 2000; Manchikanti et al., 2007; Saffier, Colombo, Brown, Mundt, & Fleming, 2007). Psychiatric comorbidity can be preexisting, or it can

Exhibit 2-10 Federal Protection of Patient Health Information

Regulation	Description
42 CFR	Applies to substance abuse treatment programs. Protects the identities and records of patients in federally assisted drug and alcohol treatment programs. With few exceptions, clinicians must obtain written consent from a patient before disclosing any information regarding his or her identity or the specific type and extent of the patient's health information, including that the patient is in an SUD treatment program.
45 CFR and Health Insurance Portability and Accountability Act (HIPAA) Privacy Rule	Applies to all clinicians. Regulates patient privacy in regard to public health. HIPAA Privacy Rule requires clinicians (or their hospitals and clinics) to safeguard information regarding patient identification and to: • Notify individuals regarding their privacy rights and how their protected health information is used or disclosed. • Adopt and implement internal privacy policies and procedures. • Train employees to understand these privacy policies and procedures as appropriate for their functions within the covered entity. • Designate individuals who are responsible for implementing privacy policies and procedures and who will receive privacy-related complaints. • Establish privacy requirements in contracts with business associates that perform covered functions. • Have in place appropriate administrative, technical, and physical safeguards to protect the privacy of health information. • Meet obligations with respect to health consumers exercising their rights under the Privacy Rule.

develop or worsen with chronic pain or SUDs. Therefore, the presence of comorbid psychiatric conditions should be assessed regularly in every patient with CNCP (see CSAT [2005b], for information on treating SUDs in people with co-occurring disorders).

Adults with chronic pain often exhibit fear about the loss of control over routine aspects of daily life; apprehension that clinicians will view their pain reports as exaggerated, imaginary, or contrived; and catastrophic thinking (hopelessness based on a conviction that things are worse than they really are). However, the distress that frequently accompanies CNCP may or may not signal a psychiatric disorder, so the clinician should try to make the distinction. Nevertheless, the decision to treat is based on the patient's level of suffering and not on whether the symptoms reach the threshold for a DSM-IV-TR diagnosis. It is often difficult to differentiate a substance-induced condition from a primary psychiatric disorder, and evaluation of symptoms over time may be necessary. Where indicated, refer patients to a mental health provider. Exhibit 2-11 identifies instruments to assess distress, anxiety, fear, and depression. Information on obtaining these instruments is in Appendix B.

Anxiety

Anxiety is common among people with CNCP and a current SUD, and it may persist in some people recovering from SUD. It is frequently associated with depression but can be present without it. Patients who have CNCP, especially those with a history of trauma, have increased rates of both anxiety symptoms and anxiety disorders (Dersh et al., 2002).

The presence of an anxiety disorder has a negative effect on treatment of CNCP. Anxiety contributes to patient suffering and can make patients less able to participate in their pain management. Treating anxiety lowers pain

scores, reduces the need for analgesics, and improves quality of life.

Depression

Patients who have CNCP and comorbid depression tend to:

- Have high pain scores.
- Feel less in control of their lives.
- Use passive–avoidant coping strategies.
- Adhere less to treatment plans than patients who are not depressed.
- Have greater interference from pain, including more pain behaviors observed by others.
- Respond less well to pain treatment, unless depression is addressed.

Clinical depression has been shown to worsen other medical illnesses, interfere with their ongoing management, and amplify their detrimental effects on health-related quality of life (Cassano & Fava, 2002; Gaynes, Burns, Tweed, & Erickson, 2002). For these reasons, depression should be treated. It may be difficult to determine whether a patient's negative affect represents clinical depression or the psychological distress of chronic pain, an SUD, or other medical conditions. Sleep apnea, hypothyroidism, and hypogonadism can present as depression. Hypogonadism is particularly relevant because it can result from prolonged exposure to opioids.

Post-Traumatic Stress Disorder

CNCP and PTSD frequently co-occur; Asmundson and colleagues (2002) report that PTSD symptoms are especially common in patients who have CNCP who have high pain scores, high pain affect, and high pain interference. Otis and colleagues (2003) recommend that patients presenting with either condition be assessed for both.

Exhibit 2-11 Tools To Assess Emotional Distress, Anxiety, Pain-Related Fear, and Depression

Tool	Purpose	Format	Administration Time
Beck Depression Inventory	Measures depression	21 items	10 minutes
Brief Patient Health Questionnaire	Measures depression, panic, stress, and women's health issues	9 items on depression, 1–5 items on panic, 13 items on stress, and 6 items on women's health	Varies
Center for Epidemiologic Studies Depression Scale	Measures how a patient has felt and behaved in past week	20 items	5–10 minutes
Geriatric Depression Scale	Seeks yes/no responses to measure depression in older adults	Short form: 15 items Long form: 30 items	5–10 minutes
Profile of Chronic Pain: Screen	Measures pain severity, interference, and emotional burden	15 items	5 minutes
Clinician Administered PTSD Scale	Assesses for PTSD symptoms, the effect of symptoms on individual's life, and the severity of symptoms	30 items	45 minutes or more
Davidson Trauma Scale	Measures frequency and severity of PTSD symptoms	17 items	10 minutes
Posttraumatic Diagnostic Scale	Assesses for PTSD symptoms and severity of symptoms	49 items	10-15 minutes
State-Trait Anxiety Inventory	Measures current anxiety and propensity for anxiety	40 items Self-administered	10–20 minutes
Tampa Scale for Kinesiophobia	Measures pain-related fear of movement; may predict disability	17 items Self-administered	5 minutes

Symptoms for CNCP and PTSD often overlap (Asmundson et al., 2002). These include anxiety, hyperarousal, avoidance behavior, emotional lability, and elevated somatic focus. Both conditions are also characterized by hypervigilance, attentional bias, stress response, and pain amplification.

Symptoms may be mutually reinforcing. For example, if CNCP resulted from a trauma, the pain may trigger flashbacks.

Somatization

Somatization refers to inordinate preoccupation with and communication about physical symptoms. Although a diagnosis of somatization disorder is rare in patients who have chronic pain, multiple pain complaints are almost always present in somatization disorder. Many patients who have multiple unexplained symptoms have subsyndromal forms of somatization disorder.

This may be categorized as *undifferentiated somatoform disorder*. When psychological factors are thought to contribute to a pain syndrome, patients may be diagnosed with *pain disorder with psychological factors* or *pain disorder with both psychological factors and a general medical condition*. Patients who have chronic pain and medically unexplained symptoms are at risk for iatrogenic consequences of unneeded diagnostic tests, medications, and surgery.

Suicide

Studies show an association between CNCP and suicidal ideation and suicide attempts that is not explained by the presence of co-occurring SUDs (Braden & Sullivan, 2008) or co-occurring mental disorders (Braden & Sullivan, 2008; Ratcliffe, Enns, Belik, & Sareen, 2008; Scott et al., 2010; Tang & Crane, 2006). In their review of 12 articles on suicide (including suicidal ideation and suicide attempts) and CNCP, Tang & Crane (2006) found that the risk for suicide "appeared to be at least doubled" in patients who experienced CNPC (p. 575). (See CSAT [2009a], for information on addressing suicidal thoughts and behaviors in substance abuse treatment).

Assessing Ability To Cope With Chronic Pain

Coping and anxiety are closely related, from a clinical viewpoint. The patient who has CNCP may have anxiety because of maladaptive coping skills, for example. The concept of acceptance has been studied in CNCP. This concept refers to the patient's belief that there is more to life than pain, that being completely free of pain is unrealistic, and that activities should be pursued, even at the price of some increase in pain (Risdon, Eccleston, Crombez, & McCracken, 2003). Patients who have high levels of acceptance report lower pain intensity, less pain-related anxiety and avoidance, less depression, less physical and psychosocial disability, more daily uptime, and better work status than do patients who have not accepted pain.

Patients who have chronic pain who score high on measures of self-efficacy or have an internal locus of control report lower levels of pain, higher pain thresholds, increased exercise performance, and more positive coping efforts (Asghari, Julaeiha, & Godarsi, 2008; Barry, Guo, Kerns, Duong, & Reid, 2003). Exhibit 2-12 lists tools to assess coping skills. Information on obtaining these instruments is provided in Appendix B.

Exhibit 2-12 Tools To Assess Coping

Tool	Purpose	Format	Administration Time
Chronic Pain Acceptance Questionnaire	Assesses willingness to experience pain and engage in activities	20 items Self-administered	5 minutes
Fear-Avoidance Beliefs Questionnaire	Assesses patients who have chronic low-back pain	16 items Self-administered	10 minutes

Evaluating Risk of Developing Problematic Opioid Use

When any patient with a behavioral health disorder is considered for opioid therapy for CNCP, the clinician must carefully weigh the risks and benefits of opioid use. Risk assessment is made over time and may change over the course of treatment (Gourlay & Heit, 2009). A patient's risk level is a matter of clinical judgment. Exhibit 2-13 presents one risk assessment schema. All patients who have SUD histories have some risk, which in many cases can be safely managed. However, in some patients, the risks of opioid use are so great and the likely benefit so small that they should not be treated with chronic opioid therapies.

Screening tools may be one element of a risk assessment. Two commonly used screening tools are the Screener and Opioid Assessment for Patients with Pain–Revised (SOAPP–R) and the Opioid Risk Tool (ORT). Both can be helpful for identifying patients at risk, but neither has been fully validated. Chapter 4 describes tools for assessing patients who have already begun opioid therapy.

Screener and Opioid Assessment for Patients with Pain–Revised

SOAPP–R can predict which patients who have CNCP are at high risk for problems with chronic opioid therapy (Exhibit 2-14) (Butler, Fernandez, Benoit, Budman, & Jamison, 2008). It is a self-administered questionnaire answered on a 5-point scale ranging from 0 (never) to 4 (very often). The numeric ratings are added; a score of 18 or higher suggests the patient is at high risk for problems with chronic opioid therapy.

Opioid Risk Tool

Opioid Risk Tool (ORT; Webster & Webster, 2005) identifies patients at risk for aberrant drug-related behaviors (ADRBs) if prescribed opioids for CNCP (Exhibit 2-15). Like SOAPP-R, ORT may help clinicians decide which patients may require close monitoring if opioids are prescribed for them. Most patients who have CNCP and histories of behavioral health disorders are likely to have elevated scores, indicating a high level of risk on opioid therapy.

Exhibit 2-13 Risk of Patient's Developing Problematic Opioid Use

Risk	Characteristics of Patient
Low	No history of substance abuse
	Minimal, if any, risk factors
Medium	History of non-opioid SUD
	Family history of substance abuse
	Personal or family history of mental illness
	History of nonadherence to scheduled medication therapy
	Poorly characterized pain problem
	History of injection-related diseases
	History of multiple unexplained medical events (e.g., trauma, burns)
High	Active SUD
	History of prescription opioid abuse
	Patient previously assigned to medium risk exhibiting aberrant behaviors

Analgesic Research, personal communication, October 30, 2009.

Exhibit 2-14 SOAPP–R Questions

1. How often do you have mood swings?
2. How often have you felt a need for higher doses of medication to treat your pain?
3. How often have you felt impatient with your doctors?
4. How often have you felt that things are just too overwhelming that you can't handle them?
5. How often is there tension in the home?
6. How often have you counted pain pills to see how many are remaining?
7. How often have you been concerned that people will judge you for taking pain medication?
8. How often do you feel bored?
9. How often have you taken more pain medication than you were supposed to?
10. How often have you worried about being left alone?
11. How often have you felt a craving for medication?
12. How often have others expressed concern over your use of medication?
13. How often have any of your close friends had a problem with alcohol or drugs?
14. How often have others told you that you have a bad temper?
15. How often have you felt consumed by the need to get pain medication?
16. How often have you run out of pain medication early?
17. How often have others kept you from getting what you deserve?
18. How often, in your lifetime, have you had legal problems or been arrested?
19. How often have you attended an Alcoholics Anonymous or Narcotics Anonymous meeting?
20. How often have you been in an argument that was so out of control that someone got hurt?
21. How often have you been sexually abused?
22. How often have others suggested that you have a drug or alcohol problem?
23. How often have you had to borrow pain medications from your family or friends?
24. How often have you been treated for an alcohol or drug problem?

Exhibit 2-15 ORT

Item	Mark Each Box That Applies	Item Score if Female	Item Score if Male
1. Family history of substance abuse			
Alcohol	☐	1	3
Illegal drugs	☐	2	3
Prescription drugs	☐	4	4
2. Personal history of substance abuse			
Alcohol	☐	3	3
Illegal drugs	☐	4	4
Prescription drugs	☐	5	5
3. Age (mark box if 16–45)	☐	1	1
4. History of preadolescent sexual abuse	☐	3	0
5. Psychological disease			
Attention deficit disorder, obsessive-compulsive disorder, bipolar, schizophrenia	☐	2	2
6. Depression	☐		
Total		______	______
Total score risk category Low risk: 0–3 Moderate risk: 4–7 High risk: ≥ 8			

Webster, L. R., & Webster, R. M. (2005). Predicting aberrant behaviors in opioid-treated patients: Preliminary validation of the Opioid Risk Tool. *Pain Medicine, 6*(6), 432–442. Reproduced with permission of Blackwell Publishing, Ltd.

Ongoing Assessment

Clinicians must assess all patients who have CNCP at regular intervals because a variety of factors can emerge that can alter treatment needs. For example, a patient may develop tolerance to a particular opioid, the underlying disease condition may change another physical or mental health problem, which might develop or worsen, or there may be changes in the patient's cognitive functioning.

Comparative data can be obtained by using the same assessment tools over time. For patients who have SUD histories or other behavioral health disorders, regular assessments should include checking for evidence of medication misuse. Chapter 4 provides a discussion on assessing and documenting the behavior of patients on opioid therapy.

The clinician should regularly:

- Assess adherence to all the recommended treatment modalities.
- Assess patient reactions to the treatment regimen.
- Determine the extent of adherence to the prescribed regimen (otherwise, the reported response may inaccurately reflect on the therapies prescribed).

- Obtain the perspectives of significant others on the patient's relief from pain, the effects of analgesia on function, and adherence to and safety with prescribed medications. (Permission to obtain collateral information is a prerequisite for prolonged opioid treatment.)

Nicholson and Passik (2007) recommend that the elements in Exhibit 2-16 be documented and kept current in a patient's record. The frequency with which these areas need to be assessed in individual patients is a matter of clinical judgment.

Treatment Setting

A clinician may conclude that optimal treatment includes more specialized care, such as that provided at a pain clinic. Where distance,

Exhibit 2-16 Elements To Document During Patient Visits

Area	Elements of Documentation
History and Physical Evaluation	History of present illness
	Pain score/intensity
	Medication history
	SUD/addiction history
	Screening tool assessments
	Medical history
	Physical examination
	Mental status/cognition
	Results of diagnostic studies
Diagnostic/Clinical Indication for Prescribing Opioids	Most probable pathological explanation of chronic pain
Treatment Plan	Pharmacological treatments
	Nonpharmacological treatments (e.g., physical therapy, exercise, behavioral therapy, lifestyle changes)
	Treatment goals and anticipated time course
	Adherence measures (e.g., urine drug testing, pill counts)
Informed Consent and Agreements for Treatment	Informed consent (e.g., discussion of risks and benefits of treatment options)
	Agreement specifying patient's responsibilities and clinic policies
Periodic Review	Pain score/intensity
	Physical, occupational, and overall function; family and social relationships; and mood and sleep patterns
	Side effects (including severity)
	ADRBs
	Medication
	Mental status/cognitive changes
Consultations and referrals	As appropriate to provide comprehensive care

Adapted from Nicholson & Passik, 2007.

costs, or other factors prohibit such a referral, the clinician must be resourceful, perhaps combining various local resources and support groups or suggesting specific electronic resources. Chapter 5 provides more details.

The vast majority of chronic pain syndromes (e.g., lumbago, osteoarthritis) in patients who do not have major psychopathology or histories of SUDs (excluding tobacco) are managed by primary care physicians. When the pain syndrome is atypical, or when there is comorbid psychiatric illness or SUD history, specialty consultation may be indicated. In the presence of current or past SUD, addiction-ology consultation may be necessary before instituting chronic therapy with scheduled medications.

Key Points

- Patients should receive a comprehensive initial assessment.
- It is important to discover the cause of a patient's chronic pain; however, clinicians should not assume a patient is disingenuous if the cause is not discovered.
- The patient's personal and family substance use histories and current substance use patterns should be assessed.
- It is crucial to obtain collateral information on the patient's pain level and functioning, as well as SUD status.
- Comorbid psychological disorders should be assessed and treated.
- Assessment of the patient with co-occurring chronic pain and SUD or other behavioral health disorders should be ongoing.

3 Chronic Pain Management

Overview of Pain Management

Chronic noncancer pain (CNCP) is a major challenge for clinicians as well as for the patients who suffer from it. The complete elimination of pain is rarely obtainable for any substantial period. Therefore, patients and clinicians should discuss treatment goals that include reducing pain, maximizing function, and improving quality of life. The best outcomes can be achieved when chronic pain management addresses co-occurring mental disorders (e.g., depression, anxiety) and when it incorporates suitable nonpharmacologic and complementary therapies for symptom management. Exhibit 3-1 presents the consensus panel's recommended strategy for treating CNCP in adults who have or are in recovery from a substance use disorder (SUD).

The Treatment Team

Chronic pain management is often complex and time consuming. It can be particularly challenging and stressful for clinicians working without input from other clinicians. The effectiveness of multiple interventions is augmented when all medical and behavioral health-care professionals involved collaborate as a team (Sanders, Harden, & Vicente, 2005). A multidisciplinary team approach provides a breadth of perspectives and skills that can enhance outcomes and reduce stress on individual providers. Although it is ideal when all relevant providers work within the same system and under the same roof, often a collaborative team must be coordinated across a community. This combined effort requires identification of a designated lead care coordinator and a good system of communication among team members and the patient. A treatment team can include the following professionals:

- Primary care provider
- Addiction specialist
- Pain clinician

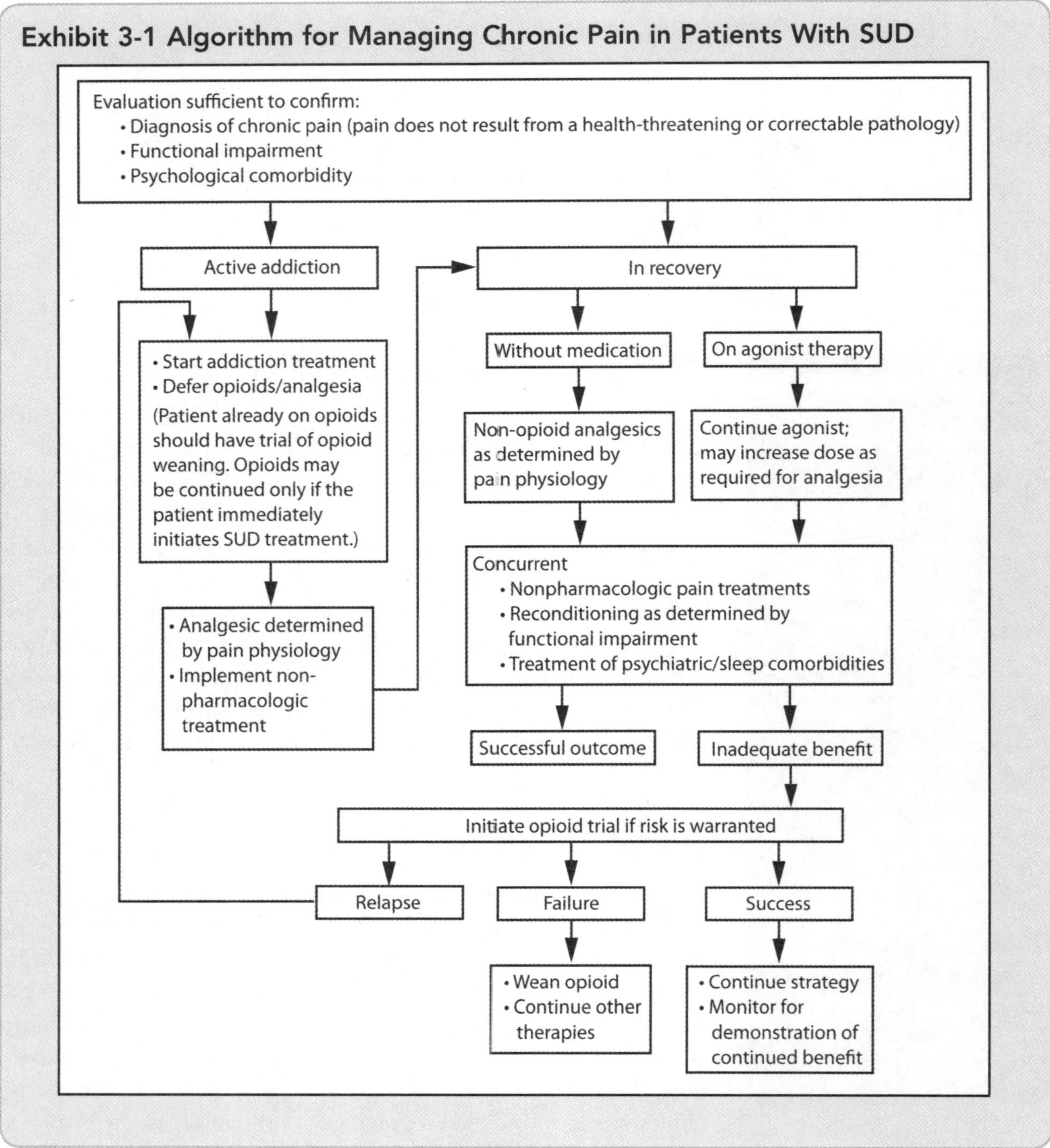

Exhibit 3-1 Algorithm for Managing Chronic Pain in Patients With SUD

- Nurse
- Pharmacist
- Psychiatrist
- Psychologist
- Other behavioral health treatment specialists (e.g., social worker, marriage and family therapist, counselor)
- Physical or occupational therapists

Addiction specialists, in particular, can make significant contributions to the management of chronic pain in patients who have SUDs. They can:

- Put safeguards in place to help patients take opioids appropriately.
- Reinforce behavioral and self-care components of pain management.

- Work with patients to reduce stress.
- Assess patients' recovery support system.
- Identify relapse.

When the addiction specialist is the prescriber of analgesics, medical responsibilities (e.g., prescribing of analgesics, physical therapy, orthotics) should be coordinated with the clinician responsible for other components of pain treatment. In some States, consultation with an addiction specialist is required before scheduled medications can be prescribed on a long-term basis to patients who have SUD histories. State laws, regulations, and policies are available at http://www.painpolicy.wisc.edu/.

The more complicated the case, the more beneficial a team approach becomes. However, many clinicians will have to treat complex patients who have little or no outside resources.

Treating Patients in Recovery

A thorough patient assessment (see Chapter 2) provides information that allows the clinician to judge the stability of a patient's recovery from an SUD. Goals for treating CNCP in patients who are in long-term recovery or whose SUD is in the distant past are as follows:

- Treat CNCP with non-opioid analgesics as determined by pathophysiology.
- Recommend or prescribe nonpharmacological therapies (e.g., cognitive–behavioral therapy [CBT], exercises to decrease pain and improve function).
- Treat comorbidities.
- Assess treatment outcomes.
- Initiate opioid therapy only if the potential benefits outweigh risk and only for as long as it is unequivocally beneficial to the patient.

Non-Opioid Analgesics

Non-opioid pharmacological options include acetaminophen and nonsteroidal anti-inflammatory drugs (NSAIDs), as well as *adjuvant medications*—so called because they originally were developed for other purposes but have analgesic properties for certain conditions. The primary adjuvant analgesics are antidepressants and anticonvulsants. Exhibit 3-2 presents a summary of these analgesics as they pertain to patients who have SUDs.

Benzodiazepines

Researchers disagree on the beneficial and harmful effects of benzodiazepines and benzodiazepine receptor agonists on chronic pain. Several studies demonstrate increased pain

Exhibit 3-2 Summary of Non-Opioid Analgesics

Analgesic	Addictive	Notes
Acetaminophen	No	Should normally not exceed 4 g/day; in adults with hepatic disease, the maximum dose is 2 g/day. Potentiates analgesia without potentiating respiratory and sedative side effects.
NSAIDs	No	Are used to relieve numerous types of pain, especially bone, dental, and inflammatory, and enhance opioid analgesia. May cause gastrointestinal bleeding and renal insufficiency.
Serotonin-Norepinephrine Reuptake Inhibitors (SNRIs)	No	Are used to relieve several nonstructural types of pain (e.g., migraine, fibromyalgia, low back pain) and probably others.

Exhibit 3-2 Summary of Non-Opioid Analgesics (continued)

Analgesic	Addictive	Notes
Tricyclic Antidepressants	No	Have demonstrated efficacy in migraine prophylaxis, fibromyalgia, many neuropathic pains, vulvodynia, and functional bowel disorders. Watch for anticholinergic side effects and orthostatic hypotension (fall risk in older people).
Anticonvulsants	No	Some have demonstrated efficacy in relieving fibromyalgia, migraine prophylaxis, and neuropathic pains.
Topical Analgesics	No	Comprise several unrelated substances (e.g., NSAIDs, capsaicin, local anesthetics). Work locally, not systemically, and therefore usually have minimal systemic side effects.
Antipsychotics	No	Have no demonstrated analgesic effect, except to abort migraine/cluster headache. Risks include extrapyramidal reactions and metabolic syndrome.
Muscle Relaxants	Carisoprodol (Soma) is addictive. Some others have significant abuse potential.	Have not been shown to be effective beyond the acute period. Some potentiate opioids and are not recommended.
Benzodiazepines	Yes	Not recommended (see discussion).
Cannabinoids	Yes	Not recommended (see discussion).

with benzodiazepines or reduced pain following benzodiazepine antagonist use (Ciccone et al., 2000; Gear et al., 1997; Nemmani & Mogil, 2003; Pakulska & Czarnecka, 2001). All benzodiazepines have side effects, including impaired coordination, reduced memory, and addiction liability. For the following reasons, the consensus panel concludes that benzodiazepines have no role in the treatment of CNCP in patients who have comorbid SUD, beyond very short-term, closely supervised treatment of acute anxiety states:

- Guidelines from the American Psychiatric Association (2006) and the United Kingdom's National Institute for Health and Clinical Excellence (Hughes et al., 2004) caution that benzodiazepines are not first-line medications.

- Excellent options to benzodiazepines for treating anxiety exist (see Treating Psychiatric Comorbidities, below).

- Anxiolytic use in adults with CNCP is often protracted.

- Benzodiazepines pose significant risk for addiction relapse and functional impairment.

The consensus panel recommends that clinicians treat comorbid anxiety and insomnia with antidepressants or anticonvulsants. Some antidepressants (e.g., trazodone, mirtazapine, amitriptyline, doxepin) may be useful sleep aids. Benzodiazepine weaning can be done in consultation with a psychiatrist or SUD treatment provider (see Center for Substance Abuse Treatment [CSAT], 2006).

Cannabinoids

At least two types of cannabinoid receptors are present in the human nervous system, and they interact with systems relevant to pain perception, including the serotonergic and dopaminergic systems. Cannabinoids are anti-inflammatory and increase levels of endogenous opioids. They inhibit glutamatergic transmission and antagonize the N-methyl-D-aspartate (NMDA) glutamate receptor, both of which actions would be expected to inhibit pain (Burns & Ineck, 2006; McCarberg, 2006).

The primary psychoactive chemical in marijuana responsible for its abuse potential is Δ9 tetrahydrocannabinol (THC). Synthetic THC (Marinol) is approved in the United States for chemotherapy-induced nausea and AIDS-induced anorexia. Sativex, a mixture of THC and cannabidiol, is an oromucosal spray that spares the lungs the toxicity of drugs and smoke. It is analgesic in neuropathic pain and is approved in Canada for the pain of multiple sclerosis. Nabilone is a synthetic drug similar to THC. Its reported analgesic effects were determined to be weaker than codeine in a controlled study of neuropathic pain (Frank, Serpell, Hughes, Matthews, & Kapur, 2008).

Although it is reasonable to conclude that modulating the human cannabinoid system shows promise for treating pain, there is no reason to believe that inhaled smoke is an acceptable delivery mode. The consensus panel does not recommend smoked marijuana for treating CNCP.

Nonpharmacological Treatments

An approach to pain management that integrates evidence-based pharmacological and nonpharmacological treatments can ease pain and reduce reliance on medication.

Nonpharmacological treatments for CNCP (Hart, 2008; Simpson, 2006):

- Pose no risk of relapse.
- May be more consistent with the recovering patient's values and preferences than pharmacological treatments, especially opioid interventions.
- May reduce pain and improve quality of life in some patients who have CNCP.
- Should be included in most pain treatment plans.

Common nonpharmacological therapies for CNCP include:

- Therapeutic exercise.
- Physical therapy (PT).
- Cognitive–behavioral therapy (CBT).
- Complementary and alternative medicine (CAM; e.g., chiropractic therapy, massage therapy, acupuncture, mind–body therapies, relaxation strategies).

Appendix D provides information on how to find qualified practitioners who provide CAM.

Therapeutic Exercise

A number of practitioners, including physicians, chiropractors, and physical therapists, frequently include exercise instruction and supervised exercise components in CNCP treatment. Therapeutic exercise can increase strength, aerobic capacity, balance, and flexibility; improve posture; and enhance general well-being. Fitness can be an antidote to the sense of helplessness and personal fragility experienced by many people with CNCP. Moderate evidence shows that exercise alleviates low back pain, neck pain, fibromyalgia, and other conditions. Furthermore, exercise reduces anxiety and depression. Limited evidence suggests that exercise benefits individuals undergoing SUD treatment (Weinstock, Barry, & Petry, 2008).

Physical Therapy

PT facilitates recovery from a large variety of medical conditions, including cardiopulmonary, geriatric, pediatric, integumentary, neurologic, and orthopedic. Neurologic PT and orthopedic PT are most likely to be used to treat chronic pain. Physical therapists use various hands-on approaches to help patients increase their range of motion, strength, and functioning. They also offer training in movement and exercises that help patients feel and function better.

Many widely used interventions by physical therapists lack definitive evidence. For example, several Cochrane Collaboration reviews of a commonly used PT modality—transcutaneous electrical nerve stimulation—found inconsistent evidence of effectiveness in a variety of chronic and acute pain conditions. Despite this lack of an evidence base, PT interventions have the advantages of being nonsurgical, bringing low risk of injury or dependence, and encouraging patients' involvement in their own recovery.

Cognitive–Behavioral Therapy

Several studies have shown that CBT can help patients who have CNCP reduce pain and associated distress, disability, depression, anxiety, and catastrophizing, as well as improve coping, functioning, and sleep (McCracken, MacKichan, & Eccleston, 2007; Thorn et al., 2007; Turner, Mancl, & Aaron, 2006; Vitiello, Rybarczyk, Von Korff, & Stepanski, 2009). In addition to its salutary effects on pain syndromes, CBT also benefits people who have SUD. In a meta-analysis of 53 controlled trials of CBT for alcohol or illicit drug disorders, CBT was found to produce a small but significant benefit (Magill & Ray, 2009).

Complementary and Alternative Medicine

CAM includes health systems, practices, and products that are not necessarily considered part of conventional medicine (National Center for Complementary and Alternative Medicine, 2007). Surveys show that 27–60 percent of chronic pain patients use CAM (Fleming, Rabago, Mundt, & Fleming, 2007; McEachrane-Gross, Liebschutz, & Berlowitz, 2006; Nayak, Matheis, Agostinelli, & Shifleft, 2001). Clinicians are urged to learn about these approaches to pain treatment not only because of their therapeutic promise, but also because many patients use CAM, raising the possibility of interactions with conventional treatments (Simpson, 2006). Exhibit 3-3 presents one way to ask patients about their use of CAM.

The evidence supporting CAM interventions for adults with comorbid CNCP and SUD is

---

Exhibit 3-3 Talking With Patients About Complementary and Alternative Medicine

Clinician	"So many of my patients use alternative medicine that I always ask about it. Are you using vitamins, herbs, acupuncture, that sort of thing for pain or for anything else?"
Patient	"Yeah. Acupuncture really helped when I was in rehab, and I still get it now and then. She does the needles and gives me Chinese herbs once in a while."
Clinician	"That's fine. If it helps, keep doing it. And when you take herbs or anything else she gives you, please tell me. I want to make sure that any herbs or medicines that you get from your acupuncturist won't interfere with the treatment that you are getting here, okay?"

ambiguous. These conditions are complex and multifactorial and, therefore, difficult to study. Many systematic reviews of CAM research note generally poor-quality reporting and heterogeneous methodology that precludes definitive evidence-based conclusions (e.g., Gagnier, van Tulder, Berman, & Bombardier, 2006). Of the CAM interventions, manual therapies are the most widely used and the most studied (Simpson, 2006). Chiropractic and massage therapies are often covered by health insurance, making these therapies accessible and compatible with conventional therapies.

Treating Psychiatric Comorbidities

Research shows well-established associations among chronic pain, SUDs, and mental disorders (e.g., depression, anxiety, post-traumatic stress disorder [PTSD], somatoform disorders) (Chelminski et al., 2005; Covington, 2007; Manchikanti et al., 2007; Saffier, Colombo, Brown, Mundt, & Fleming, 2007; Wasan et al., 2007). Psychiatric comorbidity is of special significance for two reasons. First, it is often occult. Second, untreated psychopathology is associated with poor pain treatment outcomes (Edwards et al., 2007; Williams, Jones, Shen, Robinson, & Kroenke, 2004). Therefore, management of patients who have CNCP must include intervention for co-occurring psychopathology.

Because psychiatric comorbid disorders might be preexisting, or they may develop or worsen with chronic pain or SUDs, it is important to determine the onset of psychiatric symptoms during the screening and assessment process (see Chapter 2). The psychiatric disorder needs to be included in the comprehensive treatment plan that is developed for the patient in consultation with the patient's treatment team (e.g., primary healthcare provider, substance abuse treatment counselor,

pain management provider, mental health professional). CSAT (2005b) provides detailed information on treatment strategies and models for working with individuals with a wide spectrum of psychiatric co-occurring disorders.

Benzodiazepines are generally indicated for short-term treatment of anxiety; however, anxiety associated with chronic pain commonly persists for years. Effective options include (Van Ameringen, Mancini, Pipe, & Bennett, 2004):

- Psychological and behavioral treatments.
- Selective serotonin reuptake inhibitors (SSRIs).
- SNRIs.
- Tricyclic antidepressants.
- Several anticonvulsants.

The anxiety that is often comorbid with CNCP can often be managed satisfactorily with adjuvants prescribed for the pain syndrome. Several anticonvulsants that are used for CNCP are strongly anxiolytic. In a review, Van Ameringen and colleagues (2004) found that the strongest evidence was for pregabalin (for social phobia and generalized anxiety disorder), gabapentin (for social phobia), lamotrigine (for PTSD), and valproic acid (for panic disorder). In addition, many antidepressants are effective for chronic pain and may be used to treat comorbid anxiety and depression, and both duloxetine and venlafaxine have been approved by the Food and Drug Administration for treatment of generalized anxiety disorder. Most tricyclic antidepressants are anxiolytic. Trazodone has also been found to be anxiolytic and is often used as a sedative in patients for whom benzodiazepine-like agents are undesirable. Treating comorbidities with medications that also alleviate pain can reduce polypharmacy, drug interactions, nonadherence, and, at times, financial costs.

The person who somatizes extensively may present a plethora of complaints. This situation may lead to the clinician's inappropriate discounting of all the patient's symptoms as trivial or imaginary. Clinicians should take the following steps in treating such a patient:

- Complete an inventory of all the patient's complaints.
- Emphasize history and physical examination in the evaluation.
- Validate the patient's symptoms while assuring him or her about the absence of worrisome pathology.
- Minimize expensive or invasive tests and treatments.
- Minimize use of medications with abuse liability, especially short-acting medications used as needed (PRN).
- Minimize use of passive modalities of therapy.
- Schedule regular appointments rather than PRN visits.
- Adequately treat comorbid Axis I (i.e., major psychiatric) disorders.
- Refer patients for counseling or relaxation training, as available.

Opioid Therapy

Limitations

Opioids are potent analgesics that may provide relief for many types of CNCP. However, even when effective, they have limitations, such as diminished efficacy over time (Ballantyne, 2006; Noble, Tregear, Treadwell, & Schoelles, 2008). Opioids also have adverse effects that many patients cannot tolerate (e.g., nausea, sedation, constipation). Other drawbacks include risk of addiction or addiction relapse, opioid-induced hyperalgesia (OIH), and many potential drug interactions. Serotonin syndrome is a potential adverse effect of both opioids and some medications used to treat depression, obsessive-compulsive disorder, or other behavioral health disorders. Serotonin syndrome can cause agitation, confusion, fever, and seizures, and it can be lethal if undetected or untreated. Patients who take SSRIs, SNRIs, St. John's Wort, monoamine oxidase inhibitors, lithium, or HIV medications are at increased risk of serotonin syndrome (U.S. Food and Drug Administration, 2006). In addition, patients who take opioids chronically are at increased risk of serotonin syndrome if medications such as fentanyl, meperidine, or pentazocaine are needed in emergency or surgical care settings.

Although opioids are an important treatment component for many patients, they are rarely sufficient. Chronic opioid therapy rarely shows more than one-third pain reduction in studies extending beyond 18 months, indicating that opioids are best used as one part of a multidimensional approach for most patients.

When an SUD co-occurs with CNCP, the benefits of opioids are not well established and risk of relapse is increased (Reid et al., 2002). Studies indicate that most patients who are currently addicted to prescription opioids had a prior SUD, suggesting that people in recovery are at increased risk for relapse (Potter et al., 2004; Rosenblum et al., 2003). This may be especially true when the prior SUD involved opioids, because one of the most powerful triggers for relapse is exposure to the former drug of choice (Daley et al., 2003; Gardner, 2000). Trescot and colleagues (2008) provide a detailed review.

Before Initiating Opioid Treatment

Exhibit 3-4 shows steps to take before initiating opioid therapy. Information about patient education, informed consent, and treatment plans is provided in Chapter 5.

Opioid Selection

For patients who have histories of SUDs, it is essential to minimize exposure to the euphoric effects of opioids. To reduce the likelihood of such effects, clinicians should:

- Select opioids with minimal rewarding properties (e.g., tramadol, codeine), when effective.
- Avoid prescribing supratherapeutic doses (usually demonstrated by presence of sedation, lethargy, functional impairment).
- If higher potency opioids are required, prescribe slow-onset opioids with prolonged duration of action (Mironer, Brown, Satterthwaite, Haasis, & LaTourette, 2000).

Short-acting medications have been recommended to be used preemptively before activities known to cause pain, such as PT, or for pain limited to certain times of day. There is controversy regarding the appropriateness of this suggestion for patients who have CNCP (Devulder, Jacobs, Richarz, & Wigget, 2009), and the practice is especially hazardous in people with current or past SUDs.

The route of administration may influence addiction risk, so medications that are injected or easily convertible to forms that can be injected, smoked, or snorted are often avoided in patients who have SUD. Some clinicians favor transdermal medication, with an agreement that refills are contingent on the patient's returning the used patches to demonstrate that they were not punctured, cut, or diverted.

Dose Finding

Dose finding for the patient with an SUD, especially a history of abuse of or dependence on opioids, can be complicated because of existing or rapidly developing tolerance to opioids. Also, analgesics affect individuals differently. A person who states that a particular opioid "doesn't work for me," whereas another opioid does, may be accurately reporting analgesic response.

Titration schedules appropriate for the patient with no SUD history may expose the patient in SUD recovery to a protracted period of inadequate relief. Although no schedule can be applied to everyone, a general guide is that, if low doses of opioids (other than methadone) are initiated for severe pain, they should be titrated rapidly to avoid subjecting the patient to a prolonged period of dose finding. However, if relatively high doses are initiated, titration should be slower and determined to a great extent by the half-life of the drug. For some patients, increasing the dose may lead to

---

Exhibit 3-4 Steps To Take If Opioid Therapy Is Indicated

Step 1.   Educate patient and family about treatment options, sharing the decision about the goal and expected outcome of therapy.

Step 2.   Discuss treatment agreement with the patient and family.

Step 3.   Obtain a written opioid agreement.

Step 4.   Determine and document the treatment plan.

Step 5.   Initiate a trial of opioid therapy.

Step 6.   Document details of therapy and results.

Department of Veterans Affairs & Department of Defense, 2010.

decreased functioning. It is essential that clinicians understand that dose finding for methadone can be dangerous (see Exhibit 3-5).

When an effective dose for a given patient has been determined, total opioid dose should thereafter be escalated very slowly, if at all, as tolerance develops. No study has ever shown that opioids eliminate chronic pain, other than in the very short term, so efforts to achieve a zero pain level with opioids will fail, while subjecting the patient to potentially intoxicating doses of the medication.

Relapse

For patients on chronic opioid therapy who have minor relapses and quickly regain stability, provision of substance abuse counseling, either in the medical setting or through a formal addiction program, may suffice. Opioids, if their continuation is deemed safe, must be very closely monitored, with short dispensing intervals and frequent urine drug testing. Unfortunately, many addiction treatment programs are unwilling to admit patients who are taking opioid pain medications, interpreting their prescription opioid use as a sign of active addiction.

Clinicians prescribing opioids need to establish relationships with substance abuse treatment providers who are willing to provide services for patients who need additional support in their recovery but do not require extensive services. For clinicians who treat a population with high levels of comorbid addiction, the development of onsite chemical dependence counseling services can be extremely helpful.

For relapse in patients for whom opioid addiction is a serious problem, referral to an opioid treatment program (OTP) for methadone maintenance therapy (MMT) may be the best choice. Such programs will not generally accept patients whose primary problem is pain because they do not have the resources to provide comprehensive pain management services. Patients who have chronic pain likely will not obtain adequate pain control through the single daily dose of methadone that can be provided through an OTP. Such programs may, however, be willing to collaborate in the management of patients, providing addiction treatment and allowing the prescription of additional opioids for pain management through a medical provider. Such arrangements require close communication between the OTP and the prescribing clinician so that patients who do not respond to SUD treatment can be safely withdrawn from opioids prescribed for pain. CSAT (2005a) provides more information about OTPs.

Exhibit 3-5 Methadone Titration

The titration of methadone for chronic pain is complex and potentially dangerous because methadone levels increase during the first few days of treatment. This risk is compounded by the variable half-life among individuals and the large number of drug interactions. In addition, cardiac toxicity (e.g., QT prolongation, torsade de pointes) is possible. The majority of deaths secondary to methadone occur in the first 14 days of use because:

- The initial dose is too high.
- It is titrated too quickly.
- It interacts with other drugs or medications.

Chou, Fanciullo, Fine, Adler, et al., 2009; Weschules, Baib, & Richeimer, 2008.

Another option for patients who have comorbid active addiction and CNCP is replacement of full agonist opioids with the partial opioid agonist buprenorphine (Heit, Covington, & Good, 2004; Heit & Gourlay, 2008). Benefits of this treatment include that dose escalation does not provide reinforcement and that the effects of other opioid substances may be attenuated. Buprenorphine can prevent withdrawal symptoms, allowing patients to stabilize and facilitating their progression into non-opioid and nonpharmacologic forms of pain treatment. However, buprenorphine prescribed specifically for pain is currently an off-label use (see Treating Patients in Medication-Assisted Recovery).

Opioid Discontinuation

Opioids should be discontinued if patient harm and public safety outweigh benefit. This situation may be apparent early in therapy, for example, if function is impaired by doses necessary to achieve useful analgesia. Harm also may outweigh benefit after a long period of successful treatment. Discontinuation of opioid therapy is addressed in Chapter 4.

Treating Patients in Medication-Assisted Recovery

Goals for treating CNCP in patients who are in medication-assisted recovery are the same as for patients who are in recovery without medications: reduce pain and craving and improve function. As with other patients:

- Start with recommending or prescribing nonpharmacological and non-opioid therapies.
- Treat comorbidities.
- Closely monitor treatment outcomes for evidence of benefit and harm.

Patients receiving opioid agonist treatment for addiction require special consideration when being treated for chronic pain. In these patients, the schedule and doses of opioid agonists sufficient to block withdrawal and craving are unlikely to provide adequate analgesia. Because of tolerance, a higher-than-usual dose of opioids may be needed (in addition to the maintenance dose) to provide pain relief.

Buprenorphine

Patients who have CNCP and are using sublingual buprenorphine treatment of opioid addiction pose special challenges. The drug is a partial mu agonist that binds tightly to the receptor. Because it is a partial agonist, its dose–response curve plateaus or even declines as the dose is increased. Thus, a ceiling dose limits both the available analgesia and the toxicity produced by overdose. Nevertheless, buprenorphine is an effective analgesic, and some patients who have addiction and CNCP may receive benefit for both conditions from it. To optimize analgesic efficacy, the drug should be given three times a day when pain reduction is a goal. High doses of buprenorphine can attenuate the effects of pure mu agonists given in addition to it. High doses tend to reduce the reinforcing effects of inappropriately consumed opioids but, at the same time, may reduce the effectiveness of opioids given for additional analgesia in the case of trauma or acute illness (Alford, Compton, & Samet, 2006).

Because buprenorphine has such high affinity for the mu receptor, it displaces full agonists and can induce acute opioid withdrawal; for example, if a patient on chronic methadone is given a dose of buprenorphine, acute opioid withdrawal may be precipitated (see CSAT [2004] for more information).

The use of buprenorphine for pain is off-label, albeit legal. Whereas clinicians must obtain a waiver to prescribe buprenorphine for an SUD, only a Drug Enforcement

Administration (DEA) registration is required to prescribe buprenorphine for pain. To clarify (for pharmacists) that a prescription does not require the special DEA number, it is useful to specify on the prescription that the drug is "for pain."

Methadone

Patients who have chronic pain do not obtain adequate pain control through a single daily dose of methadone because the analgesic effects of methadone are short acting in comparison with its half-life. The dosing schedule for the treatment of opioid addiction does not effectively treat pain, although the single dose often provides transient analgesia.

Methadone effects vary significantly from patient to patient, and finding a safe dose is difficult. Methadone's analgesic effects last approximately 6 hours. However, its half-life is variable and may be up to 36 hours in some patients. Pain patients may take 10 days or longer to stabilize on methadone, so the clinician must titrate very slowly and balance the risk of insufficient dosing with the life-threatening dangers of overdosing (Heit & Gourlay, 2008) (Exhibit 3-5). It is critical for the clinician to advise patients to stop methadone treatment if they become sedated.

Methadone is an especially desirable analgesic for chronic use because of its low cost and its relatively slow development of analgesic tolerance; however, it is also especially toxic because of issues of accumulation, drug interaction, and QT prolongation. For these reasons, it should be prescribed only by providers who are thoroughly familiar with it.

It is critical that patients starting methadone receive a thorough education in the dangers of inadvertent overdose with this medication. They must understand that a dose that seems initially inadequate can be toxic a few days later because of accumulation. They should be advised to keep the medication out of reach so that they cannot take a dose when sedated. Furthermore, they must be informed of the extreme danger if a child or nontolerant adult ingests their medication. Chapter 5 provides more patient education information, and CSAT (2009b) describes emerging issues in the use of methadone.

Naltrexone

Patients taking naltrexone should not be prescribed outpatient opioids for any reason. Naltrexone is a long-acting oral or injectable mu antagonist that blocks the effects of opioids. It also reduces alcohol consumption by impeding its rewarding effects. Because naltrexone displaces opioid agonists from their binding sites, opioid analgesics will not be effective in patients on naltrexone. Increasing the dose of opioids to overcome the blockade puts the patient at risk of respiratory arrest. Pain relief for these patients requires non-opioid modalities.

If patients on naltrexone require emergency opioids for acute pain, higher doses are required, which, if continued, can become toxic as naltrexone levels wane. In this situation, inpatient or prolonged emergency department monitoring is required (Covington, 2008).

Tolerance and Hyperalgesia

Tolerance develops rapidly to the sedating, euphoric, and anxiolytic effects of opioids. It develops more slowly to their analgesic effects and seldom develops to their constipating effects. Tolerance can be characterized as decreased sensitivity to opioids, whereas OIH is increased sensitivity to pain resulting from opioid use. In a clinical setting, it may be impossible to distinguish between the two conditions, and they may coexist (Angst & Clark, 2006). Tolerance can develop in chronic opioid therapy regardless of opioid type, dose, route of administration, and administration schedules (DuPen, Shen, & Ersek, 2007). Hyperalgesia has been found to result from the use of those opioids thus far studied (i.e., methadone, buprenorphine, sufentanyl, fentanyl, morphine, heroin). Patients in MMT experience analgesic tolerance and OIH. Clinical implications of these findings are unclear, as studies indicate that OIH may develop to some measures of pain (e.g., cold pressor test) and not to others (e.g., pressure) (Mao, 2002).

When patients develop tolerance to the analgesic effects of a particular opioid, either dose escalation or opioid rotation may be useful (Exhibit 3-6). *Opioid rotation*, switching from one opioid to another, is a way to exploit incomplete cross-tolerance to achieve improved analgesia without an increase in (equivalent) doses.

If a patient requests an increase in opioid dose, it is important for the clinician to try to discern whether the patient is experiencing increased pain or analgesic tolerance or is seeking some other effect (e.g., sedation, reduced anxiety). In the patient seeking sedation or reduced anxiety, a larger opioid dose provides temporary anxiolytic or sedative effects, but tolerance soon develops, necessitating another dose increase. To avoid a cycle of dose increases, the clinician should evaluate the patient's request. When nonanalgesic effects seem to be the basis for the request, alternative non-opioid medications should be provided and opioid doses should not be increased.

As with tolerance, OIH appears to require increased doses of opioids to achieve previous levels of analgesia. However, with OIH, increased doses could exacerbate pain. Treating pain with a multimodal approach—in addition to analgesics—may reduce the need for opioids, thereby decreasing the risk of tolerance and OIH.

Treating Pain in Patients Who Have Active Addiction

The presence of active addiction—whether to alcohol, opioids, or other substances—makes successful treatment of chronic pain improbable (Covington, 2008; Weaver & Schnoll, 2007). For patients who have active addiction and CNCP, it may be impossible for clinicians in the primary care setting to provide the comprehensive services necessary to treat both conditions. Specifically, an active SUD indicates that the patient should be referred for formal addiction treatment. The clinician should work closely with the patient's SUD treatment provider.

If the patient refuses the SUD referral, the clinician can use motivational interviewing techniques. CSAT (1999b) provides more information on motivational interviewing. If the patient still does not consent to addiction treatment, he or she should not be prescribed scheduled medications, except for acute pain or detoxification. CSAT (2006) provides more information on detoxification.

Exhibit 3-6 Opioid Rotation

When an opioid is ineffective, becomes ineffective, or produces intolerable side effects, it is common practice to rotate opioids. This practice is based on the observation that particular opioids affect people differently, primarily because of intraindividual and interindividual variability among opiate receptors, so-called mu-receptor polymorphism. Although most opioid analgesics are mu agonists, they affect some mu receptors differently from others. A Cochrane review (Quigley, 2004) looked at the evidence supporting the replacement of an opioid to which an individual has developed analgesic tolerance with a different opioid. The conclusion was that although evidence is scant, the practice appears to be efficacious. The most common opioid rotation, and most studied, is from morphine to methadone.

Once the patient's SUD recovery is stable, the likelihood of managing his or her pain increases. The need for formal addiction treatment often necessitates a change in the plan for opioids, by discontinuing them or by changing the treatment setting through which they are provided.

Acute Pain Episodes

When patients who have CNCP and an SUD require acute pain management, such as for postoperative pain, precautionary steps can minimize risk of relapse.

Patients in recovery may benefit from non-pharmacological pain control. Some patients in recovery from SUDs may prefer to avoid the use of any medication. Evidence shows that stress management, CBT, manual therapies, and acupuncture offer effective relief for certain types of acute pain (Hurwitz et al., 2008; Vernon, Humphreys, & Hagino, 2007).

Patients in recovery may benefit from being switched from short- to long-acting medications as quickly as appropriate (to minimize reinforcing effects). They may also benefit from bolstered recovery support during post-operative periods (Covington, 2008).

Patients on agonist therapy for addiction or pain may be continued on their current opioid or on an equivalent dose of an alternative opioid; however, this should not be expected to control acute pain, which requires supplementation with (often greater-than-usual doses of) additional opioids. In this situation, adjuvant NSAIDs may allow clinicians to provide pain relief with a reduction in opioid dosage (Mehta & Langford, 2006), and multimodal analgesia should be considered (Maheshwari, Boutary, Yun, Sirianni, & Dorr, 2006).

Patients on buprenorphine for opioid addiction may have reduced benefit from full agonist opioids used for acute pain, because the full agonist will be somewhat blocked. Non-opioid analgesics can be used, but in some cases buprenorphine will need to be discontinued so that full agonist opioids for pain can be used (Alford et al., 2006).

Patient-controlled analgesia should have relatively high bolus doses and short lockout intervals (specified intervals during which pressing the administration button results in no drug delivery), and patients should be closely monitored by medical staff. Pulse oximetry or end-tidal CO_2 monitoring may provide an additional margin of safety when high doses of opioids are required.

Patients who are dependent on opioids or sedatives (including benzodiazepines) should not be withdrawn from these medications while undergoing acute medical interventions.

Exhibit 3-7 provides a discussion of treating patients who have sickle cell disease (SCD), which brings recurring acute pain, often against a backdrop of persistent pain and hyperalgesia.

Other comorbidities that can complicate pain treatment result from other chronic illnesses. Exhibit 3-8 offers suggestions for providers for treating CNCP in patients who have HIV/AIDS.

Assessing Treatment Outcomes

Treatment of chronic pain is usually an evolving process, with medication and adjunctive therapies attempted, monitored, and adjusted or abandoned as indicated by patient response. Chapter 2 provides information about ongoing assessments.

Exhibit 3-7 Treating Patients Who Have Sickle Cell Disease

SCD is characterized by crises of acute pain, attributed to vasoocclusion, that is typically nociceptive but can be neuropathic as well. Opioids are the mainstay of treatment, although parenteral ketorolac may suffice in some crises and have opioid-sparing effects in others.

Acute pain management is critical but is often poorly conducted. At times, mutual mistrust between the patient and the clinician may lead to fears of being discounted on the part of the patient and suspicions of symptom exaggeration on the part of the clinician.

The development of CNCP further complicates the situation. When there is a clear structural explanation for pain (e.g., leg ulcers, avascular necrosis, osteomyelitis), appropriate (typically opioid) therapy is usually provided. Many patients, however, report chronic pain in the absence of detectable peripheral pathology. This pain has been attributed to central sensitization as a result of multiple episodes of severe pain. It can also result from ischemic neuropathic conditions. A small percentage of patients who have SCD develop an SUD, which adversely impact their pain reports and treatment responses.

It is generally accepted that appropriate treatment of an SCD crisis requires prompt and aggressive analgesia. Some hospitals and emergency departments keep a log of SCD patients that documents their degree of opioid tolerance, typically effective agents, and doses required so that near-immediate relief can be provided to patients presenting for care. Chronic pain with persistent tissue pathology likely requires continuation of substantial opioid doses for acceptable relief, although peripheral and adjuvant agents should be used as appropriate.

The treatment of chronic idiopathic pain in SCD often requires a multidisciplinary approach with emphasis on adjuvant analgesics and nonpharmacological therapies, including psychological therapies (Ballas, 2007).

Exhibit 3-8 Treating Patients Who Have HIV/AIDS

A vast range of pain syndromes are common in patients who have HIV/AIDS. Some are the result of HIV infection, others result from immunosuppression, and others are unrelated but comorbid with AIDS. Pain commonly results from painful neuropathy, Kaposi's sarcoma, herpes zoster, candida esophagitis, drug-induced pancreatitis, headache (including those resulting from meningitis), and numerous types of joint and myofascial pains.

Patients are often both indigent and negatively viewed by clinicians—conditions that lead to reduced access to pain care. The patients may be sick, frail, and cachectic, creating challenges in the use of pharmacotherapies. A large number of patients have a comorbid SUD, which complicates the use of opioid analgesics.

Core principles of treating CNCP, such as meticulous diagnosis of the pain mechanism and etiology and monitoring for benefits and adverse effects of treatment, and use of the World Health Organization's pain ladder (see http://www.who.int/cancer/palliative/painladder/en/index.html) for titrating analgesics are applicable in this population. However, addressing the psychological aspects of the illness, as well as functional restoration, is especially important. Nonpharmacological therapies, including PT modalities, acupuncture, biofeedback training, and hypnosis, may be helpful.

Breitbart, 2003.

Key Points

- Pain treatment goals should include improved functioning and pain reduction.
- Treatment for pain and comorbidities should be integrated.
- Non-opioid pharmacological and nonpharmacological therapies, including CAM, should be considered routine before opioid treatment is initiated.
- Opioids may be necessary and should not be ruled out based on an individual's having an SUD history.
- The decision to treat pain with opioids should be based on a careful consideration of benefits and risks.
- Addiction specialists should be part of the treatment team and should be consulted in the development of the pain treatment plan, when possible.
- A substantial percentage of patients with and without SUDs will fail to benefit from prolonged opioid therapy, in which case it should be discontinued, as with any other failed treatment.

4 Managing Addiction Risk in Patients Treated With Opioids

Promoting Adherence

Clinicians should adopt a universal precautions approach toward their patients who have chronic noncancer pain (CNCP) (Exhibit 4-1). The term *universal precautions* first emerged in the context of infectious disease treatment and referred to using infection control procedures with all patients. In the context of pain treatment, a universal precautions approach refers to a minimum standard of care applied to all patients who have CNCP, whatever their assessed risk (Gourlay, Heit, & Almahrezi, 2005). A *universal precautions approach* improves care and shows due diligence in an era of increasing illegal use of prescription opioids.

Clinicians can help patients adhere to treatment plans by:

- Employing treatment agreements.
- Regulating visit intervals.

Exhibit 4-1 Ten Steps of Universal Precautions

1. Make a diagnosis with appropriate differential.

2. Perform a psychological assessment, including risk of addictive disorders.

3. Obtain informed consent.

4. Use a treatment agreement.

5. Conduct assessments of pain level and function before and after the intervention.

6. Begin an appropriate trial of opioid therapy with or without adjunctive medications and therapies.

7. Reassess pain score and level of function.

8. Regularly assess the "4As" of pain medication (see Documenting Care, below).

9. Periodically review pain diagnosis and co-occurring conditions, including addictive disorders.

10. Document initial evaluation and followup visits.

Adapted from Gourlay et al., 2005.

- Controlling medication supply.
- Conducting urine drug testing (UDT).
- To the degree possible, including the patient's support network in monitoring efforts.

Treatment Agreements

A treatment agreement can be used to gauge and reinforce adherence to medication routines and to nonpharmacological therapies that can be critical for the patient's return to normal function. It is unlikely that a patient can follow every element of an agreement exactly at all times throughout chronic opioid therapy. The clinician's role is to note departures from the plan, to make a differential diagnosis, and to adjust the plan as needed.

Significant departures from the agreement may indicate that other members of the treatment team need to be consulted or that the patient's care should be transferred to a specialist. Any actions the patient is expected to take to return to adherence should be clearly explained. Treatment agreements are discussed at length in Chapter 5.

Visit Intervals

Patients on opioid therapy typically meet with a clinician monthly. However, patients who have histories of substance use disorders (SUDs) may require more frequent visits, such as weekly, whereas patients who are in stable recovery may be seen less often. Other factors that affect the frequency of visits include the complexity of the pain diagnosis, the status of the pain management, and the medications being prescribed.

A schedule of routine visits has advantages over sporadic appointments arranged by the patient. It encourages the patient to consider the pain a manageable condition rather than an occasionally erupting crisis. Routine also allows for close monitoring of adherence.

A patient who misses or reschedules appointments should be evaluated for relapse to an SUD.

Medication Supply

The Drug Enforcement Administration's (DEA's) "do not fill until" option allows clinicians to write a 3-month prescription that can be filled in spaced intervals (Exhibit 4-2). However, only rarely should a patient with an SUD history be seen as infrequently as every 3 months. Patients who find it difficult to adhere to treatment plans may be better served by more frequent visits during which prescriptions for smaller amounts of medication are provided. In this case, clinicians can use the "do not fill until" strategy to divide a month's supply into, for example, three 10-day prescriptions for patients who cannot handle a month's worth of medication.

Clinicians also can promote adherence through pill counts or by recruiting (with the patient's consent) a pharmacist or trusted family member to dispense medication daily. Patients who require tighter-than-weekly dispensing of medication also probably require a

---

Exhibit 4-2 Issuance of Multiple Prescriptions for Schedule II Controlled Substances

71 Federal Register 52724 allows clinicians to write multiple prescriptions—to be filled sequentially—for the same Schedule II controlled substance. The multiple prescriptions, in effect, allow a patient to receive, over time, up to a 90-day supply of that scheduled medication. But the law can also be used to write sequential prescriptions for a much shorter period, which may be appropriate for patients who benefit from tight structure. Information on other aspects of the regulation, as well as the conditions under which multiple prescriptions can be written, is at http://www.deadiversion.usdoj.gov/21cfr/cfr/2106cfrt.htm.

higher level of care and will often benefit from comanagement with an addiction specialist. Exhibit 4-3 presents a scenario regarding medication supply.

Urine Drug Testing

UDT can detect the presence of some prescribed and unprescribed substances and, therefore, can be a useful tool for improving patient care (Cone & Caplan, 2009; Heit & Gourlay, 2004; McMillin & Urry, 2007). Couto, Romney, Leider, Sharma, and Goldfarb (2009) studied data from the urine tests of more than 900,000 patients on chronic opioid treatment. The researchers found that 75 percent of the patients showed at least one sign of nonadherence to opioid regimens (e.g., having no

detectable levels of the prescribed medication in their urine, having evidence of an illicit drug). As when using other tools, the clinician must understand the limitations of UDT and interpret results in light of other clinical findings.

There are two common kinds of urine drug tests:

- Immunoassay (IA) screens, which use antibodies to detect a drug or metabolite (e.g., opioids) in urine
- Specific substance identification tests, such as GC/MS (gas chromatography/ mass spectroscopy) or HPLC (high-performance liquid chromatography), which use more sophisticated methods to detect the presence of specific substances

---

Exhibit 4-3 Talking With Patients About Medication Supply

Clinician	"I see that you are here because you ran out of your pain medication before you were due to pick up the next prescription."
Patient	"I took extra pills for a few days and now I'm out. I'm hurting more because I don't have any pills."
Clinician	"Can you tell me what happened?"
Patient	"I fell and hurt my knee, and it was really bothering me, so I took more than I usually do."
Clinician	"We have a written agreement that you'll take your medications only as prescribed."
Patient	"Yeah, but it made sense because my knee hurt so bad."
Clinician	"Knee pain is a different kind of pain, and increasing your opioid medication is not necessarily the best treatment for that. Next time, please call me first as we agreed."
Patient	"Okay, I'm sorry."
Clinician	"Whenever one of my patients breaks the agreement for any reason, I always ask for a urine sample. When did you last take your medicine?"
Patient	"I just ran out yesterday."
Clinician	"So you did not take anything else when you ran out of your prescription?"
Patient	"No! I didn't have anything else to take."
Clinician	"Okay. I'll write your prescription while you go see the nurse. If your urine sample is okay, I'll give you the prescription."

Immunoassays

Heit and Gourlay (2004) recommend testing for the following substances during routine screening: cocaine, amphetamines/methamphetamine, opioids, methadone, marijuana, and benzodiazepines.

Although IA screens are sensitive to natural opioids, they have limited ability to detect synthetic and semisynthetic opioids (e.g., hydrocodone, fentanyl, oxycodone, methadone). If an IA screen fails to detect the presence of expected opioids, the results can be confirmed with a more specific substance identification test.

IA screens can be conducted at the point of care (POC) or in a laboratory. Exhibit 4-4 lists the benefits and limitations of POC testing. POC results that are positive for an illicit substance or negative for the prescribed substance class can be verified with a confirmatory test.

Specific Substance Identification Tests

Because IA screens cannot reliably detect synthetic and semisynthetic opioids, they have limited utility for monitoring adherence to

opioid treatment. However, they can detect whether the patient on chronic opioid therapy is using, for example, marijuana, amphetamines, or cocaine. GC/MS or HPLC provide specific information about what compounds were consumed. Substance identification tests can also confirm the results of an IA screen. It is useful for clinicians prescribing chronic opioid therapy to maintain a relationship with the testing laboratory that they use so that they can communicate specific needs, such as a "no limits" test to identify small amounts of substances or specifically sought substances that may not routinely be assessed.

Urine Drug Testing Results

UDT is subject to false-positive and false-negative results. The clinician must interpret results carefully and explore the possible causes of unexpected findings before taking action. For example, prescribed medication may not show up in a UDT result because (Gourlay et al., 2006):

- The patient did not use medications or did not use them recently.

Exhibit 4-4 POC Testing Benefits and Limitations

Category	Description
Benefit	Convenient
	Fast
	Single-use kits available
	Requires little training
Limitation	Identifies only drug class or metabolite, so of limited use for adherence monitoring
	Clinicians must clearly understand cutoff scores and whether they are appropriate
	Subject to cross-reactivity (false-positive results)
	Testing instructions and cutoff scores vary
	Little quality control
	Little or no technical support

Adapted from Gourlay, Heit, & Caplan, 2006.

- The patient excretes medications or metabolites at a rate different from normal.
- The test was not sufficiently sensitive to detect the medications at the concentrations present.
- There was a clerical error.

If the results of UDT are unexpected, the clinician may want to take the following steps, recommended by Gourlay and Heit (2009):

- Contact the laboratory to confirm there was no clerical error.
- Discuss with the laboratory what type of followup test or confirmatory test should be conducted.
- Discuss the results with the patient and document the UDT results and discussion in the patient's medical record.
- Confirm disputed results with the recommended laboratory test.

An unexpected result should be discussed face-to-face with the patient (Exhibit 4-5). The presence of unprescribed or illicit substances does not render a patient's pain complaints illegitimate, but it may suggest abuse or addiction. Repeated unexpected results suggest the need for evaluation by an addiction specialist. If a patient with CNCP is diagnosed with a comorbid SUD, the patient must be willing to accept treatment for both disorders. It is reasonable for the clinician, without getting into a dispute about patients' rights to use substances or the benefits of medical marijuana, to make access to opioid therapy contingent on the patient's willingness to relinquish use of illicit substances. This can be presented simply as a way to ensure the patient's access to treatment and the clinician's continued ability to prescribe. If the person is unwilling to relinquish recreational use, the pain problem may not warrant chronic opioid therapy. If the patient is unable to relinquish the drugs, then addiction treatment is indicated (Covington, 2008).

Exhibit 4-5 Talking With Patients About Aberrant Urine Drug Testing Results

Clinician   "It seems you have not been taking your medications."

Patient A   "Yes, I have."

Clinician   "Some of it should be showing up in your urine, but it's not."

Patient A   "My husband twisted his ankle last week. I might have given him a couple of tablets."

Clinician   "You must take your medications as prescribed. That's the only way I can determine whether they are effective in treating your pain. And I need to explain to you again the harm that can result when someone else takes medication that has been prescribed to you based on your body's needs."

Clinician   "It seems you have been taking medications that I haven't prescribed."

Patient B   "No, I haven't."

Clinician   "Your last urine test was positive for benzodiazepines. Can you think of any reasons why they might have appeared?"

Patient B   "Oh, that. I was stressed out because my pain was so bad. My buddy gave me a pill."

Clinician   "There are several reasons why your pain may have gotten worse. It's really important that I know what medications you are taking and that you don't take medications that I have not prescribed for you."

Intervals for UDT depend on the degree of oversight the patient requires. The tests can be scheduled or random, depending on the patient's risk level; however, if patients are tested selectively, there is a risk of overtesting minorities and other marginalized groups and failing to detect substance abuse in patients whose ethnicity and socioeconomic status mirror those of the clinician.

Clinicians should help patients understand that UDT helps protect their recovery, their access to analgesia, and the clinician's ability to prescribe and that urine drug tests are neither punitive nor discriminatory because they are expected of all patients who receive chronic opioid therapy. Exhibit 4-6 presents sample scenarios for addressing UDT with patients.

Inclusion of Family, Friends, and Others

With support from others, the patient may be better able to comply with pain treatment. Just as important, the inclusion of others enables the clinician to obtain a clearer picture of the patient's response to treatment, including his or her ability to adhere to an opioid medication regimen, any loss of function, or development of aberrant behaviors that may indicate relapse. When a patient has a history of an SUD, it is crucial that the prescribing clinician obtain collateral information from household members, physical therapists, pharmacists, and other members of the patient's healthcare team. Given that patients may not always realize or disclose their problems with drugs, safety considerations require that prescriptions for addictive substances be contingent on the clinician's unrestricted access to collateral information.

Nonadherence

At some point in the treatment of chronic pain, patients are likely to fail to adhere to their treatment agreement. Behavior that suggests substance misuse, abuse, or addiction is known as aberrant drug-related behavior (ADRB). ADRB includes:

- Being more interested in opioids (especially immediate-release and nongeneric) than in other medications or in any other aspect of treatment.
- Taking doses larger than those prescribed or increasing dosage without consulting the clinician.
- Insisting that higher doses are needed.
- Resisting UDT, referrals to specialists, and other aspects of treatment.
- Resisting changes to opioid therapy.
- Repeatedly losing medications or prescriptions or seeking early refills.
- Making multiple phone calls about prescriptions.
- Attempting unscheduled visits, typically after office hours or when the clinician is unavailable.
- Appearing sedated.
- Misusing alcohol or using illicit drugs.
- Showing deteriorating functioning and beginning to experience adverse consequences from medications (e.g., problems at home or on the job).
- Injecting (having track marks) or snorting oral formulations.
- Obtaining medications illegally (e.g., from multiple clinicians, street dealers, family members, the Internet, forged prescriptions).

Exhibit 4-6 Talking With Patients Who Are Resistant to Urine Drug Testing

Patient A   "I can't give you a urine sample today. I just peed."

Clinician   "Not a problem. There's a water fountain in the hallway. Why don't you take in some fluids and come back when you're ready. I'll leave your refill prescription with the receptionist, and you can pick it up after we've run your screen."

Patient B   "Why do I need to give you a urine sample? Don't you trust me?"

Clinician   "The urine sample gives me a great deal of useful information about how you are using your medications and whether you are running into problems with other substances."

Patient B   "It's spying."

Clinician   "It may seem like that to you, but it's a standard part of care for all my patients. Any level of substance use can affect a patient's life and the management of the pain. Is there something we need to talk about?"

Patient C   "My problem was alcohol, doc. I was never a drug addict. You don't have to treat me like one."

Clinician   "I'm sorry if this gives you the impression that I am judging you. As your doctor, I have a job to identify and resolve any issues that may interfere with your pain treatment, sooner rather than later. The drug screen helps me do that."

Patient D   "I'm philosophically opposed to drug tests."

Clinician   "I understand. I've had other reluctant patients. Can you tell me why you feel this way?"

Patient D   "It just seems like an invasion of privacy."

Clinician   "Yes, it does. A lot of things that happen in the doctor's office can seem like an invasion of privacy. But our treatment options are limited if we can't run the test."

Patient D   "That's not fair."

Clinician   "I will still work with you as your pain doctor, no matter what the test reveals. That's fair, isn't it?"

Patient D   "I still don't want to do it."

Clinician   "I'd like to have you talk with Joe [the addiction counselor on the patient's treatment team]. He can help you sort this out."

Patient E   "I hate tests."

Clinician   "There are no passing or failing grades here. I am not going to flunk or fire you based on what I learn. In fact, we go over the results together, and we decide together how to interpret them and what to do if anything shows up unusual. How does that sound?"

Patient F   "But I gave you a urine sample last time I was here."

Clinician   "Yes, you did. Let's look at your treatment agreement. Here it is: Item 5. We agreed that you might be asked for a screen at every appointment."

- Behaving in an intimidating or threatening manner.
- Having urine drug tests that do not show the presence of prescribed opioids.
- Not adhering to nonpharmacological components of treatment.

Patient behavior is highly variable and dependent on circumstances, and the evidence base does not decisively implicate any single behavior or set of drug-related behaviors as being indicative of addiction. ADRB can be driven by other causes, including:

- Misunderstanding instructions.
- Seeking euphoria.
- Using medications to deal with fear, anger, stress, sleep problems, or other issues.
- Diverting medications for profit.
- Coping with untreated mental disorders.
- Coping with undertreated pain, also known as pseudoaddiction (Exhibit 4-7).
- General nonadherence.

ADRBs that clinicians are most likely to observe (or that patients are most likely to report) are often the behaviors that are most ambiguous (e.g., not following a medication regimen precisely, running out of a prescription early). The extreme behaviors that are easier to interpret (e.g., selling prescriptions, altering the medication's delivery mode) are ones that may elude observation during an office visit. The development of a strong therapeutic relationship facilitates these often difficult conversations when and if ADRBs occur.

Tools To Assess Aberrant Drug-Related Behaviors

Tools exist to help clinicians assess ADRBs in patients on chronic opioid therapy. However, evidence for their validity is limited (Chou, Fanciullo, Fine, Miaskowski, et al., 2009). The Addiction Behaviors Checklist (Wu et al., 2006) helps determine whether opioids have become a problem for the patient (Exhibit 4-8). It is for ongoing evaluation and can flag addiction problems as they develop. The tool can be quickly administered at each office visit; three or more "yes" responses should trigger more careful monitoring or intervention.

The Current Opioid Misuse Measure (Butler et al., 2007) asks patients about their behavior in the 30 days before the appointment (Exhibit 4-9). Butler and colleagues recommend a conservative cutoff score of 9, which yields some false-positive results, but misses fewer patients who may be misusing medications.

Exhibit 4-7 Pseudoaddiction

Patients sometimes display ADRB in response to undertreated pain. This phenomenon has been termed *pseudoaddiction* (Weissman & Haddox, 1989). It is often unclear how to determine the presence of pseudoaddiction in a patient, and the explanation of pseudoaddiction must be applied cautiously in patients who have a known SUD. Clinicians may never know with certainty what motivates ADRB in patients.

Exhibit 4-8 Addiction Behaviors Checklist

Item	Yes	No	Not Assessed
Addiction Behaviors Since Last Visit			
1. Patient used illicit drugs or evidences problem drinking			
2. Patient has hoarded medication			
3. Patient used more opioids than prescribed			
4. Patient ran out of medications early			
5. Patient has increased use of opioids			
6. Patient used analgesics PRN when prescription is for time-contingent use			
7. Patient received opioids from more than one provider			
8. Patient bought medications on the streets			
Addiction Behaviors Within Current Visit			
1. Patient appears sedated or confused (e.g., slurred speech, unresponsive)			
2. Patient expresses worries about addiction			
3. Patient expresses a strong preference for a specific type of analgesic or a specific route of administration			
4. Patient expresses concern about future availability of opioid			
5. Patient reports worsened relationships with family			
6. Patient misrepresents analgesic prescription or use			
7. Patient indicates she or he "needs" or "must have" analgesic medications			
8. Discussion of analgesic medications is the predominant issue of visit			
9. Patient exhibits lack of interest in rehabilitation or self-management			
10. Patient reports minimal/inadequate relief from opioid analgesic			
11. Patient indicates difficulty with using medication agreement			
Other			
1. Significant others express concern over patient's use of analgesics			

Reprinted from Wu et al., 2008. The Addiction Behaviors Checklist: Validation of a new clinician-based measure of inappropriate opioid use in chronic pain. *Journal of Pain and Symptom Management, 32*(4), 342–351. Adapted with permission from Elsevier.

Documenting Care

Meticulous documentation of chronic opioid therapy is essential. It is both a Federal and State requirement, and the quality of documentation can determine whether a clinician is judged to be practicing medicine or trafficking in drugs. In addition, longitudinal documentation is essential to permit a determination over time of the extent to which treatment is an asset or a liability to the patient. Documentation also provides protection for the clinician if drug enforcement authorities conduct an investigation.

The practitioner must be familiar with the requirements of the State in which he or she practices; however, generally there must be documentation of an adequate medical workup of the condition being treated, an evaluation for psychiatric comorbidity including SUD, a plan of care, amounts of scheduled medications prescribed, and instructions for use of medications. Some States require that chronic opioid therapy be used only if other treatments are ineffective or ill-advised. The University of Wisconsin's Pain and Policy Studies Group maintains a Web site that describes the regulations of different States regarding opioid prescribing (http://www.painpolicy.wisc.edu/).

Exhibit 4-9 Current Opioid Misuse Measure

(Measured on a Scale of 0=Never to 4=Very Often)

1. How often have you had trouble with thinking clearly or had memory problems?

2. How often do people complain that you are not completing necessary tasks (i.e., doing things that need to be done, such as going to class, work, or appointments)?

3. How often have you had to go to someone other than your prescribing physician to get sufficient pain relief from your medications (i.e., another doctor, the emergency room)?

4. How often have you taken your medications differently from how they are prescribed?

5. How often have you seriously thought about hurting yourself?

6. How much of your time was spent thinking about opioid medications (having enough, taking them, dosing schedule, etc.)?

7. How often have you been in an argument?

8. How often have you had trouble controlling your anger (road rage, screaming, etc.)?

9. How often have you needed to take pain medications belonging to someone else?

10. How often have you been worried about how you're handling your medications?

11. How often have others been worried about how you're handling your medications?

12. How often have you had to make an emergency phone call or show up at the clinic without an appointment?

13. How often have you gotten angry with people?

14. How often have you had to take more of your medication than prescribed?

15. How often have you borrowed pain medication from someone else?

16. How often have you used your pain medicine for symptoms other than for pain (to help you sleep, improve your mood, relieve stress, etc.)?

17. How often have you had to visit the emergency room?

Reprinted from Butler, et al., 2007. Development and validation of the Current Opioid Misuse Measure. *Pain, 130*, 144–156. Used with permission from Elsevier.

Like other treatments, opioid therapy should be continued only so long as it is effective. Many clinicians have found it useful to monitor and document opioid response using the "4As" of Passik and colleagues (2004): analgesia, activities of daily living, adverse events, and ADRBs.

Managing Difficult Conversations

Patients who have CNCP can be especially difficult to treat because their condition often eludes diagnosis and because unremitting pain itself can affect their ability to be civil. When an SUD or other co-occurring disorder is overlaid onto the pain, the likelihood of difficult behavior from the patient increases. Such a patient has complex and intense needs that are best served by a treatment team approach that allows for frequent assessment and care of the patient without overburdening any one member of the team (see Chapter 3).

The following activities can help build a therapeutic relationship between the treatment team and the patient:

- Listening actively
- Asking open-ended, nonjudgmental questions
- Restating a patient's report to make sure it has been understood
- Using clarification statements ("It sounds as if the pain is worse than usual for you")
- Demonstrating empathy
- Using feeling statements ("This must be very difficult for you")

One strategy for demonstrating empathy is to specifically acknowledge the effort required simply to cope with pain daily. The clinician should not promise overly optimistic results and should educate patients so that they form reasonable expectations about outcomes. It may also help to suggest that patients focus on improvements in functioning and avoid defining their lives by their pain.

Patients who have chronic pain, as well as those with SUDs, and perhaps especially those with both, can elicit strong negative responses from treatment providers. These negative reactions impede efforts to experience and communicate empathy. It is useful, first of all, for the clinician to recognize these reactions and to seek to understand them. Frequently, they are simply a result of the frustration attendant on treating difficult or intractable problems. They may result from feelings that the clinician is working harder for the patient's wellness than is the patient. It may help for the clinician to remind himself or herself that, no matter how lacking in motivation the patient seems, no one would ever wish for a typical life of a person with comorbid pain and addiction.

Workplace Safety

Clinicians and their patients must be protected from violence in the workplace. Clinic staff members should be encouraged to be proactive and aware of their surroundings, report suspicious activity, and use common sense to make good decisions about aggressive patients or family members. Clinicians should plan for occasional disruptive or aggressive behavior and position themselves in the examination room between the patient and the door. If a patient becomes threatening, security personnel or law enforcement may be needed. The consensus panel recommends that clinicians develop crisis management policies and plans and ensure that staff members are trained and drilled on their implementation. A plan should be developed for contacting public safety officials (discreetly, if necessary) in urgent or emergent situations. The plan should include a distress signal to alert all staff members. Contact information for public safety officials should be readily available.

Drug Diversion

Some patients sell or trade their medications, and sometimes patients give their medications to family or friends for various reasons. Medications taken by people other than for whom they are prescribed are said to be "diverted."

Unequivocal evidence of diversion is rare, although patients often acknowledge it when confronted. All members of the treatment team should be alert to the patient who:

- Is known to have contact with people with active SUDs.
- Cannot produce the remainder of a partially used prescription when asked for a pill or patch count.
- Has attempted to alter or forge prescriptions.
- Has been "doctor shopping" to obtain additional medications.
- Does not comply with the nonpharmacological components of recommended treatment.
- Strongly prefers brand name drugs or drugs with high street value.
- Fails to demonstrate the presence of prescribed opioids in appropriate UDT results.

Clinicians should know which drugs are popular in their communities and be vigilant when prescribing medications that have high street value (Exhibit 4-10). Many clinicians scrupulously avoid prescribing medications with high marketability to patients who have an addiction disorder or histories of diversion.

Clinicians should remind patients of their responsibility to protect their medications against theft and diversion (see Chapter 5). Clinicians must understand and comply with State laws regarding prescribing practices. State laws on the amount of opioids prescribed and prescription expiration may be more restrictive than Federal laws. State laws can be found at the Federation of State Medical Boards Web site (http://www.fsmb.org).

Some clinicians have inordinately restricted their prescribing because of a false belief that there has been a "crackdown" on clinicians prescribing opioids for pain. DEA's policy statement on dispensing scheduled medications for the treatment of pain, which includes a response to concerns of a crackdown, is found at http://www.deadiversion.usdoj.gov/fed_regs/notices/2006/fr09062.htm.

Strict boundaries should be placed around a patient who pushes for medications that the clinician believes are unwise choices. In this situation, the clinician's responsibility is to prescribe what is indicated and not what the patient desires. Although all diversion is unlawful, there are degrees of seriousness; for example, a patient who gives a hydrocodone tablet to a spouse who sustained a back sprain should not be treated in the same way as a "pseudo-patient" who seeks medications to resell. If diversion is suspected, treatment monitoring must be tightened. Clinicians

Exhibit 4-10 Resources for Information on Drug Use Trends

Resource	Web Site
National Institute on Drug Abuse's Community Epidemiology Work Group	http://www.drugabuse.gov/about/organization/CEWG/CEWGHome.html
Drug Abuse Warning Network	http://dawninfo.samhsa.gov
Centers for Disease Control and Prevention National Center for Health Statistics	http://www.cdc.gov/nchs

should not tolerate any serious diversion, which is a breach of trust that usually calls for cessation of opioid therapy or even ending the clinician–patient relationship. Evidence of diversion should be documented.

State prescription monitoring programs, which currently operate in 38 States, may be useful to clinicians who suspect a patient of "doctor shopping," that is, obtaining scheduled medications from multiple clinicians (Wang & Christo, 2009). Information regarding programs and a list of States that have active programs are at http://www.namsdl.org/presdrug.htm.

Discontinuation of Opioid Therapy

The best reason to discontinue opioid therapy is that the pain has resolved, but that is often not the case. Other likely reasons for discontinuation include the following:

- Opioids are no longer effective.
- Opioids no longer stabilize the patient or improve function.
- The patient loses control over the medication.
- The patient is diverting the medication.
- The patient is using alcohol, benzodiazepines, or illicit drugs.
- Adverse effects are unmanageable.

When the benefits of opioid therapy are outweighed by its harms, therapy should be discontinued. Of course, this statement applies to all medications, of whatever category. Exhibit 4-11 presents an algorithm for discontinuing chronic opioid therapy.

Patients tapering off opioids may experience both short-term withdrawal (which occurs immediately) and protracted withdrawal. Short-term withdrawal begins when the level of opioid in the blood falls below the accustomed level for the patient. It usually abates after a few days or a week (depending on the half-life of the medication). Patients may experience increased pain and withdrawal hyperalgesia. Symptoms and signs of opioid withdrawal are as follows.

Symptoms include:

- Abdominal cramps, nausea, vomiting, diarrhea
- Bone and muscle pain
- Anxiety
- Insomnia
- Increased pain sensitivity in the original painful site

Signs include:

- Tachycardia
- Hypertension
- Fever
- Mydriasis
- Hyperreflexia
- Diaphoresis
- Piloerection
- Lacrimation, yawning
- Rhinorrhea
- Myoclonus

Protracted withdrawal from opioids includes anxiety, depression, sleep disturbances, fatigue, dysphoria, and irritability, which can last for weeks or months following withdrawal from short- and long-acting opioids (Collins & Kleber, 2004; Satel, Kosten, Schuckit, & Fischman, 1993). These symptoms can be attenuated with tricyclic antidepressants, gabapentin, and other nonaddicting agents. Discomfort may develop at any time during the weaning process, so patients should be monitored until the process is complete and any symptoms addressed. Cognitive–behavioral therapy may help with cravings. Not all patients experience protracted withdrawal.

Exhibit 4-11 Exit Strategy

Determine whether risks outweigh benefits of opioid therapy in this patient, employing the following criteria:
- Inability to maintain analgesia despite reasonable dose escalation
- Intolerable side effects at the minimum dose that produces effective analgesia
- Reasonable attempts at opioid rotation unsuccessful
- Persistent noncompliance with patient treatment agreement
- Deterioration in physical, emotional, or social functioning attributed to opioid therapy

Establish collaborative relationship with patient around need for discontinuation of opioid therapy:
- Review exit criteria agreed on in patient treatment agreement
- Clarify that exit is for patient's (not doctor's) benefit
- Clarify that exiting opioid therapy is not synonymous with abandoning pain management or abandoning the patient

Patient appears to have a problem with drug addiction.	No apparent addiction problem. Patient able to cooperate with office-based taper.	Patient unable or unwilling to cooperate with outpatient taper.
Refer for addiction management or comanagement.	• Taper opioids slowly, as tolerated; typically 25% of current dose can be reduced each week. • If withdrawal discomfort persists, use clonidine or similar agents to attenuate. Doxepin or gabapentin may also be useful. • Implement non-opioid pain management strategies, including psychosocial support; cognitive-behavioral therapies; physical therapy; non-opioid analgesics; and management of insomnia, anxiety, and depression.	• Provide sufficient opioid for 1-month taper or maintenance until admission to SUD treatment. • Refer to inpatient program or comprehensive outpatient program, or similar services, as available. • If patient runs out early or fails to enter recommended treatment, then provide weekly prescriptions for doses reduced by 25% from the preceding week. • If concerned about inadvertent overdose, prescribe frequent (daily, if necessary) nonlethal amounts as tapering proceeds.

Note: This algorithm does not indicate a maximum trial dose, as none has been established by research. However, it can be said that doses above 200 mg morphine equivalents per day have not been studied systematically, and higher doses are more likely to be associated with active addiction than are lower doses (Ballantyne, 2006). Some clinicians have recommended doses up to 300 mg morphine equivalents per day. Analgesic Research, personal communication, October 30, 2009; Covington, personal communication, October 30, 2009.

For patients whose active addiction necessitates discontinuation of opioid therapy, referral for specialized addiction treatment is crucial.

There are many reasons for discontinuing scheduled medications but very few for discontinuing care of the patient. When opioids are a liability, whether because of poor analgesic efficacy or patient ADRB, the clinician should usually offer to continue to provide non-opioid therapies and treatment; that is, stopping opioids does not mean stopping treatment.

The clinician who elects to discharge a patient from his or her practice should inform the patient in writing. To avoid charges of abandonment, the clinician should provide the patient with contact information for other clinicians, along with a written tapering schedule and prescriptions for the medications that require a taper. In cases in which the clinician–patient relationship is hostile or dangerous or in which the patient presents a danger to the clinician, a letter alone can suffice.

Key Points

- Patients on chronic opioid therapy should be monitored closely for signs of benefit, harm, and ADRBs.
- All ADRBs should be documented, investigated, and acted on.
- Difficult conversations should be managed with compassion and empathy.
- Clinicians should establish and respectfully maintain strict limits with patients who insist on opioids.
- Clinicians should establish relationships with drug-testing laboratory staff and addiction specialists.
- When it is necessary to discontinue chronic opioid therapy, a conscientious tapering plan should be provided.

5 Patient Education and Treatment Agreements

The Value of Patient Education

No randomized controlled trials have specifically evaluated the effect of patient education on treatment outcomes; however, Brox and colleagues (2006) studied 60 patients who had persistent low back pain at least a year after surgery for disc herniation. Patients were randomized to receive either lumbar fusion with transpedicular screws or cognitive intervention, which consisted largely of education on back hygiene and exercises. Outcomes were essentially the same for the two groups.

The potential value of patient education is also supported by ad hoc reviews in the medical literature. For instance, *VA/DoD Clinical Practice Guideline for the Management of Opioid Therapy for Chronic Pain* (Department of Veterans Affairs & Department of Defense [VA/DoD], 2010) recommends both patient and family education, as do other pain treatment guidelines (Chou, Fanciullo, Fine, Adler, et al., 2009; Institute for Clinical Systems Improvement, 2007). Patient education is also necessary for truly informed consent. Geppert (2004, p. 163) defines informed consent as follows: "Informed consent encompasses the capacity to understand the risks, benefits, and alternatives of a treatment, to communicate a choice regarding therapy, to deliberate and reason about the consequences of the proposed medication, and to appreciate how the treatment will affect life and values." Informed consent is particularly important when clinicians are prescribing potentially addictive medications to patients who have histories of substance use disorders (SUDs) and other behavioral health disorders.

Providing culturally sensitive and linguistically appropriate education can:

- Improve adherence.
- Help the patient understand medication responses that are expected and normal and those that are of concern and warrant a phone call.
- Allay fears about particular treatments or medications.

- Increase satisfaction with treatment by promoting realistic expectations.
- Provide an opportunity to discuss any concerns.
- Strengthen the clinician–patient relationship by demonstrating respect and enhancing patient feelings of self-efficacy.
- Improve health, well-being, and outcomes.

In addition, patient education provides a forum in which clinicians can ask patients about their perceptions of their condition and explore patients' conceptions and misconceptions about their condition and its treatment. Clinicians should encourage patients to talk about their use of complementary and alternative medicine (CAM) and other non-pharmacological approaches to pain.

Providing education and soliciting questions require an initial time commitment; however, these efforts ultimately save time. A patient is less likely to make unnecessary emergency appointments when he or she clearly understands what to expect from a medication or treatment and has a specific plan for what actions to take when pain flares.

Providing Effective Education

Effective education is a process that begins at treatment initiation and continues throughout treatment. The treatment needs of patients who have chronic noncancer pain (CNCP) and SUD change over time, necessitating ongoing education. Family members, especially caregivers, frequently play important roles in pain treatment (Glajchen, 2001) and ought to be involved in educational efforts.

Educational approaches must be tailored to each patient's needs. The clinician or other members of the treatment team should develop a repertoire of educational materials and approaches to meet the differing needs of patients. Approaches should consider:

- Primary languages spoken by patients.
- Culture, gender, race/ethnicity, and age of patients.
- Resources in the local community (e.g., availability of hospitals, pharmacies).
- Educational, general literacy, and health literacy levels of patients.
- Cognitive function of patients.

Health literacy has been defined as "the capacity to obtain, process, and understand basic health information and services needed to make appropriate health decisions" (U.S. Department of Health and Human Services, 2000, p. vi). Online sources for more information and training on health literacy are in Exhibit 5-1.

Exhibit 5-1 Selected Online Sources of Information on Health Literacy

Organization	Web Site
American Medical Association Health Literacy Kit (continuing medical education credits available)	http://www.ama-assn.org/ama/pub/physician-resources.shtml
Health Resources and Services Administration (HRSA) Unified Health Communication 101: Addressing Health Literacy, Cultural Competency, and Limited English Proficiency	http://www.hrsa.gov/healthliteracy/default.htm
Center for Health Care Strategies, Inc., Health Literacy Fact Sheets	http://www.chcs.org/usr_doc/Health_Literacy_Fact_Sheets.pdf
Institute of Medicine Health Literacy: A Prescription to End Confusion	http://www.iom.edu/Reports/2004/Health-Literacy-A-Prescription-to-End-Confusion.aspx

To enhance communication with a diverse patient population, HRSA recommends that the treatment team (http://www.hrsa.gov/healthliteracy):

- Use simple language and short sentences and define technical terms.
- Supplement instruction with appropriate materials (e.g., videos, models, pictures).
- Ask patients to explain or demonstrate the clinician's instructions (teach-back method; see Exhibit 5-2).
- Ask open-ended questions that begin with "how" and "what," rather than closed-ended, yes/no questions.
- Organize information so that the most important points stand out and repeat this information.
- Consider gender; age; and the cultural, ethnic, and racial diversity of patients when selecting or designing educational materials.
- Offer assistance with completing forms.

Oral communication can be supplemented with charts, diagrams, and other visual aids. These can help patients with limited English proficiency or low-literacy skills as well as those who learn more efficiently from graphic representations. Patient education materials are available in several languages at http://www.healthinfotranslations.org. In some cases, a translator may be necessary.

Communication also can be enhanced by using the teach-back method (or "interactive communication loop") (Schillinger et al., 2003). This method can be effectively used with any patient but may be particularly helpful with those who have low general knowledge or health literacy or who are in early recovery from an SUD.

The method involves the clinician's explaining or demonstrating an instruction or concept to the patient, then asking the patient to repeat the instruction or information back in his or her own words (not verbatim) or to repeat the demonstration. When asking, the clinician takes responsibility for any misunderstanding (e.g., "I want to be sure I explained this well enough"). If the patient cannot demonstrate or does not appear to understand the instruction, the clinician tries again. This is repeated until the patient clearly understands what he or she is expected to do. Exhibit 5-2 offers a sample teach-back dialog.

Exhibit 5-2 Talking With Patients Following a Teach-Back Approach

Clinician	"You're going to take this medicine, the green pill, two times each day: once in the morning, once at bedtime. Now, to be sure I explained this well enough, please tell me in your own words how you'll take this medicine."
Patient	"Uh … I'm going to take two green pills in the morning and two green pills when I go to bed."
Clinician	"Well, no. I'm sorry I wasn't clear. You will take one pill in the morning and one pill before you go to bed. Now, tell me again how you'll take this medication."
Patient	"Okay, I think I've got it now. I'm going to take one pill in the morning and one pill before bed."
Clinician	"That's it. I am also going to write the instructions out for you. I also want to go over some information on when to call me. I am going to attach this information to your prescription. Please call me with any questions or concerns."

Take-home handouts and pamphlets may aid recall and provide additional information. Clinicians can prepare their own handouts, but many can be found online. University medical centers and government Web sites are a good source of reliable patient education resources, and pharmaceutical companies almost always offer patient education sheets on specific medications. Professional associations also often have useful materials, such as patient communication aids (e.g., Pain Log, Quality of Life Scale). Clinicians should review these documents for appropriateness, print them out, go over them with patients, and allow patients to take them home.

The Internet as a Source of Patient Education

Many patients use the Internet as a source of health information. Although the Internet can be a useful adjunct to in-office education, it is also a source of much misinformation and marketing disguised as education.

Clinicians can offer guidance and recommend Web sites with reliable content on chronic pain management and SUDs. Exhibit 5-3 lists a few such Web sites with information on chronic pain, and Exhibit 5-4 lists Web sites with information on SUDs.

Exhibit 5-3 Reliable Web Sites With Information on Chronic Pain and Pain Treatment

Organization	Web Site
Aetna Intellihealth	http://www.intellihealth.com
Agency for Healthcare Research and Quality	http://www.ahrq.gov
American Academy of Family Physicians	http://www.familydoctor.org
American Academy of Pain Medicine	http://www.painmed.org
American Cancer Society	http://www.cancer.org
American Chronic Pain Association	http://www.theacpa.org
American Pain Foundation	http://www.painfoundation.org
American Pain Society	http://www.ampainsoc.org
American Society of Anesthesiologists	http://www.asahq.org
Arthritis Foundation	http://www.arthritis.org
Breastcancer.org	http://www.breastcancer.org
Emerging Solutions in Pain	http://www.emergingsolutionsinpain.com
Komen Foundation	http://ww5.komen.org
MedicineNet, Inc.	http://www.medicinenet.com
National Cancer Institute	http://www.cancer.gov (click PDQ, physician data query)
National Institutes of Health	http://www.nih.gov
National Pain Foundation	http://www.nationalpainfoundation.org
Veterans Affairs	http://www.va.gov/painmanagement
WebMd	http://www.webmd.com

Exhibit 5-4 Reliable Web Sites With Information on Substance Use Disorders

Organization	Web Site
National Institute on Alcohol Abuse and Alcoholism	http://www.niaaa.nih.gov
National Institute on Drug Abuse	http://www.drugabuse.gov
National Library of Medicine	http://www.nlm.nih.gov
Parents. TheAntiDrug.com	http://www.theantidrug.com
Partnership for a Drug-Free America	http://www.drugfree.org
Substance Abuse and Mental Health Services Administration	http://www.samhsa.gov

Education Content

General Information

The specifics of patient education vary from patient to patient and over time. However, general content areas for patient education include information about:

- The patient's condition and the nature of the patient's chronic pain.
- Treatments available, including nonpharmacological options.
- The risks and benefits of treatment options.
- How and when to take medications.
- How to keep medications safely away from children (out of reach or locked up).
- The patient's responsibility for keeping track of medications and not losing them or giving them to others.
- Any medication interactions.
- Common side effects of medication, their expected duration, and ways to manage them (e.g., a high-fiber diet to manage constipation common to opioid use).
- Warnings and potential adverse events associated with medications and other treatments.
- Pros and cons of CAM.
- Risks to pregnant and lactating women.

- The degree of pain relief the patient can realistically expect from a treatment.
- How to use treatment apparatus (e.g., transcutaneous electrical nerve stimulation machine).
- How best to use the Internet to find information and sources of support.
- Under what conditions the patient should immediately call the clinician or go to the emergency department.
- How to deal with episodes of acute pain (e.g., from surgery or trauma), as well as flareup pain.

Patients may benefit from referrals to psychologists for assistance in basic coping skills and to physical therapists and other professionals (Naliboff, Wu, & Pham, 2006) for therapies that can be used in place of or in addition to medication (e.g., meditation, relaxation, stretching, distraction).

Opioid Information

Use of opioids requires additional educational efforts. To give informed consent, patients must understand the expected benefits as well as the uncertainties of chronic opioid therapy. Specifically, they must understand that excellent analgesia can almost always be provided by starting opioids; however, long-term studies are limited and often of poor quality. They suggest that benefit diminishes with time; after 1½ years, about one-half of patients dropped out of opioid therapy because of side effects or the therapy's loss of efficacy. Those continuing to take opioids had about 30-percent pain reduction (Kalso, Edwards, Moore, & McQuay, 2004).

Patients must also understand the risks of therapy, which include overdose (by patient, others, pets), constipation, sedation, and hormone changes, and the hazards of combining opioids with sedating drugs or alcohol. Finally, they should understand that tolerance and physical dependence are expected consequences of extended therapy, that these conditions do not necessarily indicate the presence of an addictive disorder, but that they do require that arrangements be made to prevent abrupt withdrawal when either the patient or clinician is out of town or the clinician is otherwise unavailable. Policies of the clinician or program (e.g., requirements for urine drug testing, responses to lost or stolen prescription reports, early refill requests) should be communicated in advance.

In addition, patients need to understand (VA/DoD, 2010):

- The titration process, how soon the patient can expect maximum effectiveness, and why taking medications exactly as prescribed is important to the titration process.
- The risks of discontinuing the medication abruptly (e.g., withdrawal symptoms).
- How medication will be safely discontinued (e.g., tapering, managing withdrawal symptoms).
- That drowsiness is a common side effect during titration and that patients should not try to drive or operate heavy machinery until drowsiness is cleared.
- How to discuss pain therapy, analgesic needs, and recovery status with other health professionals (e.g., dentists, anesthesiologists). (See Exhibit 5-5.)

Exhibit 5-5 Talking With Patients Before Surgery

Anesthesiologist: "I understand you are scheduled for knee-replacement surgery. Is there anything else you would like me to know about your health?"

Patient: "Yes, I was addicted to Vicodin many years ago. I still have chronic low back pain even though I had a laminectomy and fusion about 5 years ago. The pain following surgery was terrible. I do not want to go through that again. Doctor, I do not want to suffer. "

Anesthesiologist: "I see. Please tell me more. I want to make sure that I have all the information so that your surgeon and I can develop a pain management plan to address your concerns."

Patient: "I take buprenorphine. It works very well for me."

Anesthesiologist: "Thank you for sharing this information. Knee surgery patients can feel a lot of pain afterward. Since we have 2 days before your surgery, we should be sure the doctor prescribing your buprenorphine knows that you are scheduled for surgery. She may want to change your buprenorphine dose. Also, we will consider using other pain medications or techniques to manage your pain. We'll need to closely monitor you and assess your pain. We will also want to be sure that your support people are aware of the plan and are available to help you. May I contact your doctor and nurses?"

Patients also need to know about legal and regulatory issues (VA/DoD, 2010), including:

- The legal responsibilities of the clinician.
- That it is illegal to give away, trade, share, or sell prescription opioids.
- The potential effect of regulatory issues on occupation, lifestyle, and use (e.g., pilots, commercial drivers; Chou and colleagues [2009] provide more information).
- The responsibility of the patient to report stolen medications both to the police and to the clinician.

Methadone Maintenance Therapy Information

Patients on methadone maintenance therapy (MMT) for opioid dependence need to understand how pain treatment will affect their MMT and vice versa. Patients also need to understand that long-term use of opioids can bring tolerance, may cause them to become more sensitive to pain (to have opioid-induced hyperalgesia), and can cause the opioids to become ineffective over time. (Chapter 3 provides more information on opioid-induced hyperalgesia.)

In general, when patients receiving MMT have inadequate pain control, options include non-opioid therapies and dividing the daily methadone dose into three-times-a-day dosing. If a decision is made to increase the dose of methadone by the pain-treating clinician, it should be done only in concert with the MMT program. The patient must be monitored for continued participation in an aggressive recovery program and for evidence that the increased dose of methadone leads to demonstrable reductions in pain or improvements in function.

Treatment Agreements

As with patient education, opioid treatment agreements (contracts) have had no randomized controlled trials that have specifically evaluated their effect on treatment outcomes. Such agreements are, however, recommended in clinical guidelines and are frequently used in practice. Although written agreements specific to prescribed opioids are most frequently discussed, agreements can be used for other treatment modalities (e.g., exercise regimens).

Disagreement exists about the use of agreements when prescribing opioids (Heit, 2003). Some guidelines recommend opioid agreements only when the patient has or is at risk for an SUD. Others are concerned that "opioid contracts may diminish patient autonomy; autonomy and adherence may sometimes represent conflicting values in chronic opioid therapy" (Arnold, Han, & Seltzer, 2006, p. 294).

These concerns can be mitigated somewhat by the way in which treatment agreements are established. Patients can be informed that treatment agreements are mutually agreed-on plans and courses of action. Providing education on options and involving the patient in planning and writing treatment agreements can preserve patient autonomy while establishing necessary guidelines. Arnold and colleagues (2006) suggest that, if a clinician chooses to use an opioid agreement, it should:

- Use neutral, nonconfrontational language.
- Be written so that the patient can understand it.
- Emphasize opioids as a part of a comprehensive pain management plan that also includes physical therapy, counseling, and other medications for co-occurring disorders, as needed.
- Emphasize the clinician's responsibility to work with the patient to alleviate his or her symptoms.

- Explain that the agreement protects the patient's access to scheduled medications and protects the clinician's license to prescribe them.
- Describe behaviors that are incompatible with chronic opioid therapy (e.g., getting prescriptions from other clinicians, losing medications).
- Describe the actions the clinician may take in response to these behaviors up to and including cessation of opioid prescribing.

As when treating all patients, the clinician can assess the ability of the patient with or in recovery from an SUD to make an informed decision (Longo, Parren, Johnson, & Kinsey, 2000). If the clinician becomes aware of limitations, he or she can (in addition to or instead of having a written agreement) involve the patient's family in treatment, with the patient's permission (Geppert, 2004).

Treatment agreements vary considerably from practice to practice and from patient to patient. However, some common elements of agreements include the following (Fishman, 2007; Heit, 2003; Jacobson & Mann, 2004; VA/DoD, 2010; Ziegler, 2007):

- Timeframe of the agreement
- Goals of therapy
- Risks and benefits of chronic opioid therapy
- Requirement for obtaining prescriptions from a single clinician and a named pharmacy
- Activities for pain management (e.g., exercise, CAM)
- Risk and benefit statement, including lists of possible side effects

- Proscription against changing medication dosage without permission
- Schedule for regular medical visits for evaluation of the agreed-on treatment
- Requirement of complete, honest self-report of pain relief, side effects, and function at each medical visit
- Limits on medication refills
- Limits on replacing lost medications or prescriptions
- Consent for random urine drug testing and other specified testing
- Required pill counts
- Consent for appropriate release of information (e.g., from family members, other clinicians, counselors, substance abuse treatment programs)
- Participation in agreed-on SUD recovery activities (e.g., treatment, continuing care, mutual-help groups)
- Requirements of the clinician
- Participation in agreed-on psychiatric treatment activities
- Possible consequences of not following the treatment agreement

A useful treatment agreement should be revised as the patient's needs and circumstances change. An opioid agreement by American Academy of Pain Management is online at http://www.aapainmanage.org. Exhibit 5-6 presents another sample pain treatment agreement for a woman in recovery from an SUD.

Exhibit 5-6 Sample Pain Treatment Agreement

Patient: Irene Simpson   **Doctor:** Dr. Miller   **Date:** 1-19-10

This treatment plan has been developed to manage neck pain and tension headaches. It is open to changes when both the doctor and I agree that the changes are in my best interest and are likely to improve my pain management or overall health. A primary goal of the plan is to protect my recovery from addiction.

1. My daily medications:
    gabapentin, 1,200 mg three times daily.
    duloxetine, 90 mg every morning.
    topiramate, 100 mg at bedtime.

2. At the first indication of a headache, I will take ibuprofen (600 mg).

3. If possible, I will lie down in a darkened room with an ice pack to my neck and shoulders for 15 to 20 minutes to give the medication time to work; if the headache is still present in 30 minutes, I will take acetaminophen (500 mg). Use of opioid medications can be considered if this plan is unsuccessful. However, under no circumstances will I seek these medications from other doctors, friends, or the Internet. Instead, I will discuss my cravings and sense that the plan is not working with Dr. Miller, Joan Small, and my sponsor.

4. I will see Dr. Wong weekly or as recommended for acupuncture treatments.

5. I will walk 15 to 30 minutes daily.

6. I will attend the pain management group with Joan weekly and see Joan for individual sessions as indicated.

7. I will obtain all prescriptions for headache or other pain and for addiction recovery from Dr. Miller, and I will fill all prescriptions at the Main Street Pharmacy.

8. I will not visit other physicians or the emergency department without first talking to Dr. Miller or to the doctor who is covering for him.

9. I will attend my home group (Tuesday Night Women's Group) weekly, plus two other weekly Narcotics Anonymous (NA) meetings of my choice; I will talk with my sponsor at least weekly and will call her when I feel despondent or have cravings to drink or take opioid pills.

10. My daily meditation will focus on removing myself from conflicts where I do not have a direct role to play. I will try to remind myself when "I don't have a horse in this race" at work or at home.

Important Phone Numbers:

Dr. Miller's Office 222-3800

Dr. Miller's Answering Service 222-9000

Main Street Pharmacy 380-2000

Joan Small's Office 380-2132

NA Hotline 234-0081

Abby (sponsor) 382-9970

Patient: ________________________________ Doctor:_______________________________ Date:__________

Sample Pain Treatment Agreement ©MediCom Worldwide, Inc., 101 Washington St., Morrisville, PA 19067. Ziegler, P. **Treating Chronic Pain in the Shadow of Addiction. Monograph 2007. Available at:** http://www.emergingsolutionsinpain.com/index.php?option=com_continued&view=dispfm&Itemid=280&course=42

Key Points

- Patient education is necessary for informed consent, and it equips patients to take an active role in their pain management.
- Education must be tailored to the individual patient. More research is needed on tailoring education to patients who have CNCP.
- Clinicians should take time to educate their patients and make sure patients understand how to help themselves.
- People learn in different ways; clinicians should have a variety of learning materials at their disposal.
- Treatment agreements document the treatment plan and the responsibilities of the patient and the clinician.

Appendix A—Bibliography

Alford, D. P., Compton, P., & Samet, J. H. (2006). Acute pain management for patients receiving maintenance methadone or buprenorphine therapy. *Annals of Internal Medicine, 144*(2), 127–134.

American Academy of Pain Medicine, American Pain Society, & American Society of Addiction Medicine Committee on Pain and Addiction. (2001). *Definitions related to the use of opioids for the treatment of pain* (consensus statement). Retrieved February 10, 2011, from http://www.ampainsoc.org/advocacy/opioids2.htm

American Psychiatric Association. (2000). *Diagnostic and statistical manual of mental disorders* (4th ed., text revision). Washington, DC: Author.

American Psychiatric Association. (2006). *American Psychiatric Association practice guidelines for the treatment of psychiatric disorders: Compendium.* Arlington, VA: Author.

Angst, M. S., & Clark, D. J. (2006). Opioid-induced hyperalgesia: A qualitative systematic review. *Anesthesiology, 104,* 570–587.

Arnold, R. M., Han, P. K., & Seltzer, D. (2006). Opioid contracts in chronic nonmalignant pain management: Objectives and uncertainties. *American Journal of Medicine, 119*(4), 292–296.

Asghari, A., Julaeiha, S., & Godarsi, M. (2008). Disability and depression in patients with chronic pain: Pain or pain related beliefs? *Archives of Iranian Medicine, 11*(3), 263–269.

Asmundson, G. J., Coons, M. J., Taylor, S., & Katz, J. (2002). PTSD and the experience of pain: Research and clinical implications of Shared Vulnerability and Mutual Maintenance models. *Canadian Journal of Psychiatry, 47*(10), 930–937.

Ballantyne, J. C. (2006). Opioids for chronic nonterminal pain. *Southern Medical Journal, 99*(11), 1245–1255.

Ballas, S. K. (2007). Current issues in sickle cell pain and its management. *Hematology*, 97–105. Retrieved February 10, 2011, from http://asheducationbook.hematologylibrary.org/cgi/content/full/2007/1/97

Barry, L. C., Guo, Z., Kerns, R. D., Duong, B. D., & Reid, M. C. (2003). Functional self-efficacy and pain-related disability among older veterans with chronic pain in a primary care setting. *Pain, 104*(1–2), 131–137.

Bird, J. (2003). Selection of pain measurement tools. *Nursing Standard, 18*(13), 33–39.

Braden, J. B., & Sullilvan, M. D. (2008). Suicidal thoughts and behavior among adults with self-reported pain conditions in the National Comorbidity Survey Replication. *Journal of Pain, 9*(12), 1106–1115.

Breitbart, W. (2003). Pain. In J. F. O'Neill, P. A. Selwyn, & H. Schietinger (Eds.), *A clinical guide to supportive and palliative care for HIV/AIDS* (pp. 85–122). Rockville, MD: Health Resources and Services Administration. Retrieved February 10, 2011, from ftp://ftp.hrsa.gov//hab/pall/chap4.PDF

Brookoff, D. (2005). Chronic pain as a disease: The pathophysiology of disordered pain. In B. McCarberg & S. D. Passik (Eds.), *Expert guide to pain management* (pp. 1–33). Philadelphia: American College of Physicians.

Brox, J. I., Reikerås, O., Nygaard, Ø., Sørensen, R., Indahl, A., Holm, I., et al., (2006). Lumbar instrumented fusion compared with cognitive intervention and exercises in patients with chronic back pain after previous surgery for disc herniation: A prospective randomized controlled study. *Pain, 122*(1–2), 145–155.

Brunton, S. (2004). Approach to assessment and diagnosis of chronic pain. *Journal of Family Practice, 53*(Suppl 10), S3–S10.

Burns, T. L., & Ineck, J. R. (2006). Cannabinoid analgesia as a potential new therapeutic option in the treatment of chronic pain. *Annals of Pharmacotherapy, 40*, 251–260.

Butler, S. F., Budman, S. H., Fernandez, K. C., Houle, B., Benoit, C., Katz, N., et al. (2007). Development and validation of the Current Opioid Misuse Measure. *Pain, 130*, 144–156.

Butler, S. F., Fernandez, K., Benoit, C., Budman, S. H., & Jamison, R. N. (2008). Validation of the revised screener and opioid assessment for patients with pain. *Journal of Pain, 9*, 360–372.

Campbell, J. N., & Meyer, R. A. (2006). Mechanisms of neuropathic pain. *Neuron, 52*(1), 77–92. Retrieved February 10, 2011, from http://www.pubmedcentral.nih.gov/articlerender.fcgi?tool=pubmed&pubmedid=17015228

Cassano, P., & Fava, M. (2002). Depression and public health: An overview. *Journal of Psychosomatic Research, 53*(4), 849–857.

Center for Substance Abuse Treatment. (1999a). *Brief interventions and brief therapies for substance abuse*. Treatment Improvement Protocol 34. HHS Publication No. (SMA) 99-3353. Rockville, MD: Substance Abuse and Mental Health Services Administration.

Center for Substance Abuse Treatment. (1999b). *Enhancing motivation for change in substance abuse treatment.* Treatment Improvement Protocol 35. HHS Publication No. (SMA) 99-3354. Rockville, MD: Substance Abuse and Mental Health Services Administration.

Center for Substance Abuse Treatment. (2004). *Clinical guidelines for the use of buprenorphine in the treatment of opioid addiction.* Treatment Improvement Protocol 40. HHS Publication No. (SMA) 04-3939. Rockville, MD: Substance Abuse and Mental Health Services Administration.

Center for Substance Abuse Treatment. (2005a). *Medication-assisted treatment for opioid addiction in opioid treatment programs.* Treatment Improvement Protocol 43. HHS Publication No. (SMA) 05-4048. Rockville, MD: Substance Abuse and Mental Health Services Administration.

Center for Substance Abuse Treatment. (2005b). *Substance abuse treatment for persons with co-occurring disorders.* Treatment Improvement Protocol 42. HHS Publication No. (SMA) 05-3922. Rockville, MD: Substance Abuse and Mental Health Services Administration.

Center for Substance Abuse Treatment. (2006). *Detoxification and substance abuse treatment.* Treatment Improvement Protocol 45. HHS Publication No. (SMA) 06-4131. Rockville, MD: Substance Abuse and Mental Health Services Administration.

Center for Substance Abuse Treatment. (2007). *National Summit on Recovery Conference report, September 28–29, 2005.* HHS Publication No. (SMA) 07-4276. Rockville, MD: Substance Abuse and Mental Health Services Administration.

Center for Substance Abuse Treatment. (2009a). *Addressing suicidal thoughts and behaviors in substance abuse treatment.* Treatment Improvement Protocol 50. HHS Publication No. (SMA) 09-4381. Rockville, MD: Substance Abuse and Mental Health Services Administration.

Center for Substance Abuse Treatment. (2009b). Emerging issues in the use of methadone. HHS Publication No. (SMA) 09-4368. *Substance Abuse Treatment Advisory, 8*(1).

Chelminski, P. R., Ives, T. J., Felix, K. M., Prakken, S. D., Miller, T. M., Perhac, J. S., et al. (2005). A primary care, multi-disciplinary disease management program for opioid-treated patients with chronic non-cancer pain and a high burden of psychiatric comorbidity. *BMC Health Services Research, 5*(1), 3.

Chou, R., Fanciullo, G. J., Fine, P. G., Adler, J. A., Ballantyne, J. C., Davies, P., et al., for the American Pain Society–American Academy of Pain Medicine Opioids Guidelines Panel. (2009). Clinical guidelines. *Journal of Pain, 10*(2), 113–130.

Chou, R., Fanciullo, G. J., Fine, P. G., Miaskowski, C., Passik, S. D., & Portenoy, R. K. (2009). Opioids for chronic noncancer pain: Prediction and identification of aberrant drug-related behaviors—A review of the evidence for an American Pain Society and American Academy of Pain Medicine clinical practice guideline. *Journal of Pain, 10*(2), 131–146.

Ciccone, D. S., Just, N., Bandilla, E. B., Reimer, E., Ilbeigi, M. S., & Wu, W. (2000). Psychological correlates of opioid use in patients with chronic nonmalignant pain: A preliminary test of the downhill spiral hypothesis. *Journal of Pain Symptom Management, 20*(3), 180–192.

Collins, E. D., & Kleber, H. D. (2004). Opioids: Detoxification. In M. Galanter & H. D. Kleber (Eds.), *Textbook of substance abuse treatment* (pp. 265–289). Arlington, VA: American Psychiatric Publishing.

Compton, P., & Gebhart, G. F. (2003). The neurophysiology of pain in addiction. In A. S. Graham, T. K. Schultz, M. F. Mayo-Smith, R. K. Ries, & B. B. Wilford (Eds.), *Principles of addiction medicine* (3rd ed., pp. 901–917). Chevy Chase, MD: American Society of Addiction Medicine.

Cone, E. J., & Caplan, Y. H. (2009). Urine toxicology testing in chronic pain management. *Postgraduate Medicine, 121*(4), 91–102.

Couto, J. E., Romney, M. C., Leider, H. L., Sharma, S., & Goldfarb, N. I. (2009). High rates of inappropriate drug use in the chronic pain population. *Population Health Management, 12*(4), 185–190.

Covington, E. C. (2007). Chronic pain management in spine disorders. *Neurologic Clinics, 25*(2), 539–566.

Covington, E. C. (2008). Pain and addictive disorder: Challenge and opportunity. In H. T. Benzon, J. P. Rathmell, C. L. Wu, D. C. Turk, & C. E. Argoff (Eds.), *Raj's practical management of pain* (4th ed., pp. 793–808). Philadelphia: Elsevier Mosby.

Daley, D. C., Marlatt, G. A., & Spotts, C. E. (2003). Relapse prevention: Clinical models and intervention strategies. In A. W. Graham, T. K. Schultz, M. F. Mayo-Smith, R. K. Ries, & B. B. Wilford (Eds.), *Principles of addiction medicine* (3rd ed., pp. 467–485). Chevy Chase, MD: American Society of Addiction Medicine.

Dennis, M. L., Foss, M. A., & Scott, C. K. (2007). An eight-year perspective on the relationship between the duration of abstinence and other aspects of recovery. *Evaluation Review, 31*(6), 585–612.

Department of Veterans Affairs & Department of Defense. (2003). *VHA pain outcomes toolkit.* Washington, DC: Veterans Health Administration, U.S. Department of Defense.

Department of Veterans Affairs & Department of Defense, Management of Opioid Therapy for Chronic Pain Working Group. (2010). VA/DoD clinical practice guideline for management of opioid therapy for chronic pain. Washington, DC: Department of Veterans Affairs, U.S. Department of Defense.

Dersh, J., Polatin, P. B., & Gatchel, R. J. (2002). Chronic pain and psychopathology: Research findings and theoretical considerations. *Psychosomatic Medicine, 64,* 773–786.

Devulder, J., Jacobs, A., Richarz, U., & Wiggett, H. (2009). Impact of opioid rescue medication for breakthrough pain on the efficacy and tolerability of long-acting opioids in patients with chronic non-malignant pain. *British Journal of Anaesthesia, 103*(4), 576–585.

Dick, B. D., & Rashiq, S. (2007). Disruption of attention and working memory traces in individuals with chronic pain. *Anesthesia & Analgesia, 104,* 1223–1229.

DuPen, A., Shen, D., & Ersek, M. (2007). Mechanisms of opioid-induced tolerance and hyperalgesia. *Pain Management Nursing, 8*(3), 113–121.

Edlund, M. J., Sullivan, M., Steffick, D., Harris, K. M., & Wells, K. B. (2007). Do users of regularly prescribed opioids have higher rates of substance use problems than nonusers? *Pain Medicine, 8*(8), 647–656.

Edwards, R. R., Klick, B., Buenaver, L., Max, M. B., Haythornthwaite, J. A., Keller, R. B., et al. (2007). Symptoms of distress as prospective predictors of pain-related sciatica treatment outcomes. *Pain, 130*(1–2), 47–55.

Fishman, S. (2007). *Responsible opioid prescribing: A physician's guide.* Washington, DC: Waterford Life Sciences.

Fleming, S., Rabago, D. P., Mundt, M. P., & Fleming, M. F. (2007). CAM therapies among primary care patients using opioid therapy. *BMC Complementary and Alternative Medicine, 7,* 15.

Folstein, M. F., & Folstein, S. E. (2010). *Mini-Mental State Examination* (2nd ed.). Lutz, FL: PAR, Inc.

Frank, B., Serpell, M. G., Hughes, J., Matthews, J. N., & Kapur, D. (2008). Comparison of analgesic effects and patient tolerability of nabilone and dihydrocodeine for chronic neuropathic pain: Randomised, crossover, double blind study. *British Medical Journal, 336*(7637), 199–201.

Gagnier, J. J., van Tulder, M. W., Berman, B. M., & Bombardier, C. (2006) *Herbal medicine for low back pain.* Retrieved February 10, 2011, from http://www.cochrane.org/reviews/en/ab004504.html

Gardner, E. L. (2000). What we have learned about addiction from animal models of drug self-administration. *American Journal on Addictions, 9*(4), 285–313.

Gaynes, B. N., Burns, B. J., Tweed, D. L., & Erickson, P. (2002). Depression and health-related quality of life. *Journal of Nervous and Mental Disease, 190*(12), 799–806.

Gear, R. W., Miaskowski, C., Heller, P. H., Paul, S. M., Gordon, N. C., & Levine, J. D. (1997). Benzodiazepine mediated antagonism of opioid analgesia. *Pain, 71*(1), 25–29.

Geppert, C. M. (2004). To help and not to harm: Ethical issues in the treatment of chronic pain in patients with substance use disorders. *Advances in Psychosomatic Medicine, 25,* 151–171.

Glajchen, M. (2001). Chronic pain: Treatment barriers and strategies for clinical practice. *Journal of the American Board of Family Practice, 14*(3), 211–218.

Gourlay, D. L., & Heit, H. A. (2009). Compliance monitoring in chronic pain management. In J. C. Ballantyne, J. P. Rathmell, & S. M. Fishman (Eds.), *Bonica's management of pain* (4th ed.). Baltimore: Lippincott Williams & Wilkins.

Gourlay, D. L., Heit, H. A., & Almahrezi, A. (2005). Universal precautions in pain medicine: A rational approach to the treatment of chronic pain. *Pain Medicine, 6*(2), 107–112.

Gourlay, D. L., Heit, H. A., & Caplan, Y. (2006). *Urine drug testing in clinical practice* (3rd ed.). San Francisco: California Academy of Family Physicians.

Green, C. R., Baker, T. A., Smith, E. M., & Sato, Y. (2003). The effect of race in older adults presenting for chronic pain management: A comparative study of black and white Americans. *Journal of Pain, 4*(2), 82–90.

Haefeli, M., & Elfering, A. (2006). Pain assessment. *European Spine Journal, 15*(Suppl 1), S17–S24.

Hart, J. (2008). Complementary therapies for chronic pain management. *Alternative and Complementary Therapies, 14*(2), 64–68.

Hart, R. P., Martelli, M. F., & Zasler, N. D. (2000). Chronic pain and neuropsychological functioning. *Neuropsychology Review,10*(3),131–149.

Hart, R. P., Wade, J. B., & Martelli, M. F. (2003). Cognitive impairment in patients with chronic pain: The significance of stress. *Current Pain and Headache Reports, 7*(2), 116–126.

Heit, H. A. (2003). Creating and implementing opioid agreements. *Disease Management Digest, 7*(1), 2–3.

Heit, H. A., Covington, E., & Good, P. A. (2004). Dear DEA. *Pain Medicine, 5*(3), 303–308.

Heit, H. A., & Gourlay, D. L. (2004). Urine drug testing in pain medicine. *Journal of Pain and Symptom Management, 27*(3), 260–267.

Heit, H. A., & Gourlay, D. L. (2008). Buprenorphine: New tricks with an old molecule for pain management. *Clinical Journal of Pain, 24*(2), 93–97.

Hughes, I., Kelly, J., Kosky, N., Lear, G., Owens, L., Ratcliffe, J., et al. (2004). *Clinical guidelines and evidence review for panic disorder and generalised anxiety disorder.* Sheffield, UK: University of Sheffield/London: National Collaborating Centre for Primary Care.

Hurwitz, E. L., Carragee, E. J., van der Velde, G., Carroll, L. J., Nordin, M., Guzman, J., et al. (2008). Treatment for neck pain: Noninvasive interventions. *Journal of Manipulative and Physiological Therapeutics, 33*(45), S123–S152.

Institute for Clinical Systems Improvement. (2007). *Assessment and management of chronic pain* (2nd ed.). Bloomington, MN: Author.

International Association for the Study of Pain/Subcommittee on Taxonomy. (1986). Classification of chronic pain syndromes and definitions of pain terms. *Pain, 3*(Suppl 3), S1–S226.

Jacobson, P. L., & Mann, J. D. (2004). The valid informed consent-treatment contract in chronic non-cancer pain: Its role in reducing barriers to effective pain management. *Comprehensive Therapy, 30*(2), 101–104.

Kalso, E., Edwards, J. E., Moore, R. A., & McQuay, H. J. (2004). Opioids in chronic non-cancer pain: Systematic review of efficacy and safety. *Pain, 112*(3), 372–398.

Karoly, P., Ruehlman, L. S., Aiken, L. S., Todd, M., & Newton, C. (2006). Evaluating chronic pain impact among patients in primary care: Further validation of a brief assessment instrument. *Pain Medicine, 7*(4), 289–298.

Katz, N. P., & Fanciullo, G. J. (2002). Role of urine toxicology testing in the management of chronic opioid therapy. *Clinical Journal of Pain, 18*(Suppl 4), S76–S82.

Koob, G. F. (2009). Neurobiological substrates for the dark side of compulsivity in addiction. *Neuropharmacology, 56*(Suppl 1), S18–S31.

Lebovits, A. H. (2000). Approaches to psychological assessment prior to multidisciplinary chronic pain management. In *Chronic pain management: Guidelines for multidisciplinary program development* (pp. 173–187). New York: Informa.

Longo, L., Parran, T., Johnson, B., & Kinsey, W. (2000). Addiction: Part II—Identification and management of the drug-seeking patient. *American Family Physician, 61*(8), 2401–2408.

Magill, M., & Ray, L. A. (2009). Cognitive-behavioral treatment with adult alcohol and illicit drug users: A meta-analysis of randomized controlled trials. *Journal of Studies on Alcohol and Drugs, 70*(4), 516–527.

Maheshwari, A. V., Boutary, M., Yun, A. G., Sirianni, L. E., & Dorr, L. D. (2006). Multimodal analgesia without routine parenteral narcotics for total hip arthroplasty. *Clinical Orthopaedics and Related Research, 453*, 231–238.

Manchikanti, L., Giordano, J., Boswell, M. V., Fellows, B., Manchukonda, R., & Pampati, V. (2007). Psychological factors as predictors of opioid abuse and illicit drug use in chronic pain patients. *Journal of Opioid Management, 3*(2), 89–100.

Mao, J. (2002). Opioid-induced abnormal pain sensitivity: Implications in clinical opioid therapy. *Pain, 100*, 213–217.

McCabe, S. E., Cranford, J. A., & Boyd, C. J. (2006). The relationship between past-year drinking behaviors and nonmedical use of prescription drugs: Prevalence of co-occurrence in a national sample. *Drug and Alcohol Dependence, 1*(3), 281–288.

McCarberg, W. (2006). A new class of analgesics: Cannabinoids (2 parts). *Pain Medicine News, 4*(4).

McCracken, L. M., MacKichan, F., & Eccleston, C. (2007). Contextual cognitive-behavioral therapy for severely disabled chronic pain sufferers: Effectiveness and clinically significant change. *European Journal of Pain, 11*(3), 314–322.

McCracken, L. M., Vowles, V. E., & Eccleston, C. (2004). Acceptance of chronic pain: Component analysis and a revised assessment method. *Pain, 107*(1–2), 159–166.

McEachrane-Gross, F., Liebschutz, J. M., & Berlowitz, D. (2006). Use of selected complementary and alternative medicine (CAM) treatments in veterans with cancer or chronic pain: A cross-sectional survey. *BMC Complementary and Alternative Medicine, 6*, 34.

McMillin, G., & Urry, F. (2007). *Drug testing guide for chronic pain management services*. Salt Lake City, UT: ARUP Laboratories.

Mehta, V., & Langford, R. M. (2006). Acute pain management for opioid dependent patients. *Anaesthesia, 61*(3), 269–276.

Mironer, Y. E., Brown, C., Satterthwaite, J., Haasis, J., & LaTourette, P. (2000). *Relative misuse potential of different opioids: A large pain clinic experience*. Atlanta, GA: American Pain Society.

Naliboff, B. D., Wu, S. M., & Pham, Q. (2006). Clinical considerations in the treatment of chronic pain with opiates. *Journal of Clinical Psychology, 62*(11), 1397–1408.

National Association of State Alcohol and Drug Abuse Directors. (2006). *Current research on screening and brief intervention and implications for State alcohol and other drug (AOD) systems*. Retrieved February 10, 2011, from: http://www.nasadad.org/resource.php?base_id=788

National Center for Complementary and Alternative Medicine. (2007). *What is Complementary and Alternative Medicine?* Retrieved February 10, 2011, from http://nccam.nih.gov/health/whatiscam

National Center for Health Statistics. (2006). *Health, United States, with chartbook on trends in the health of Americans*. Washington, DC: U.S. Government Printing Office.

National Institute on Alcohol Abuse and Alcoholism. (2005). *Helping patients who drink too much: A clinician's guide*. Bethesda, MD: Author. Retrieved February 10, 2011, from http://pubs.niaaa.nih.gov/publications/Practitioner/CliniciansGuide2005/clinicians_guide.htm

National Institute on Drug Abuse. (2007). *Drugs, brains, and behavior: The science of addiction*. NIH Publication No. 07-5605. Bethesda, MD: Author. Retrieved February 10, 2011, from http://www.drugabuse.gov/scienceofaddiction

National Institute on Drug Abuse. (2009). *Principles of addiction treatment*. NIH Publication No. 09-4180. Bethesda, MD: Author.

Nayak, S., Matheis, R. J., Agostinelli, S., & Shifleft, S. C. (2001). The use of complementary and alternative therapies for chronic pain following spinal cord injury: A pilot survey. *Journal of Spinal Cord Medicine, 24*(1), 54–62.

Nemmani, K. V., & Mogil, J. S. (2003). Serotonin-GABA interactions in the modulation of mu- and kappa-opioid analgesia. *Neuropharmacology, 44*(3), 304–310.

Nestler, E. J. (2005). The neurobiology of cocaine addiction. *Science and Practice Perspectives, 3*(1), 4–12.

Nicholson, B., & Passik, S. D. (2007). Management of chronic noncancer pain in the primary care setting. *Southern Medicine Journal, 100*(10), 1028–1036.

Noble, M., Tregear, S. J., Treadwell, J. R., & Schoelles, K. (2008). Long-term opioid therapy for chronic noncancer pain: A systematic review and meta-analysis of efficacy and safety. *Journal of Pain Symptom Management, 35*(2), 214–228.

Otis, J. D., Keane, T. M., & Kerns, R. D. (2003). An examination of the relationship between chronic pain and post-traumatic stress disorder. *Journal of Rehabilitation Research and Development, 40*(5), 397–406.

Pakulska, W., & Czarnecka, E. (2001). Effect of diazepam and midazolam on the antinociceptive effect of morphine, metamizol and indomethacin in mice. *Pharmazie, 56*(1), 89–91.

Passik, S. D., & Kirsh, K. L. (2004). Opioid therapy in patients with a history of substance abuse. *CNS Drugs, 18*(1), 13–25.

Passik, S. D., Kirsh, K. L., Whitcomb, L., Portenoy, R. K., Katz, N. P., Kleinman, L., et al. (2004). A new tool to assess and document pain outcomes in chronic pain patients receiving opioid therapy. *Clinical Therapeutics, 26*(4), 552–561.

Peles, E., Schreiber, S., Gordon, J., & Adelson, M. (2005). Significantly higher methadone dose for methadone maintenance treatment (MMT) patients with chronic pain. *Pain, 113*, 340–346.

Pence, L. B., Thorn, B. E., Jensen, M. P., & Romano, J. M. (2008). Examination of perceived spouse responses to patient well and pain behavior in patients with headache. *Clinical Journal of Pain, 24*(8), 654–661.

Portenoy, R. K., Ugarte, C., Fuller, I., & Haas, G. (2004). Population-based survey of pain in the United States: Differences among White, African American, and Hispanic subjects. *Journal of Pain, 5*(6), 317–328.

Potter, J. S., Hennessy, G., Borrow, J. A., Greenfield, S. F., & Weiss, R. D. (2004). Substance use histories in patients seeking treatment for controlled-release oxycodone dependence. *Drug and Alcohol Dependence, 76*(2), 213–215.

Potter, J. S., Shiffman, S. J., & Weiss, R. D. (2008). Chronic pain severity in opioid-dependent patients. *American Journal of Drug and Alcohol Abuse, 34*(1), 101–107.

Quigley, C. (2004). Opioid switching to improve pain relief and drug tolerability. *Cochrane Database of Systematic Reviews, 3*, CD004847.

Ratcliffe, G. E., Enns, M. W., Belik, S.-L., & Sareen, J. (2008). Chronic pain conditions and suicidal ideation and suicide attempts: An epidemiologic perspective. *Clinical Journal of Pain, 24*, 204–210.

Reid, M. C., Engles-Horton, L. L., Weber, M. B., Kerns, R. D., Rogers, E. L., & O'Connor, P. G. (2002). Use of opioid medications for chronic noncancer pain syndromes in primary care. *Journal of General Internal Medicine, 17*(3), 173–179.

Risdon, A., Eccleston, C., Crombez, G., & McCracken, L. (2003). How can we learn to live with pain? A Q-methodological analysis of the diverse understandings of acceptance of chronic pain. *Social Science and Medicine, 56*(2), 375–386.

Rosenblum, A., Joseph, H., Fong, C., Kipnis, S., Cleland, C., & Portenoy, R. (2003). Prevalence and characteristics of chronic pain among chemically dependent patients in methadone maintenance and residential treatment facilities. *Journal of the American Medical Association, 289*(18), 2370–2378.

Rupp, T., & Delaney, K. A. (2004). Inadequate analgesia in emergency medicine. *Annals of Emergency Medicine, 43*(4), 494–503.

Saffier, K., Colombo, C., Brown, D., Mundt, M., & Fleming, M. (2007). Addiction Severity Index in a chronic pain sample receiving opioid therapy. *Journal of Substance Abuse Treatment, 33*, 303–311.

Sanders, S. H., Harden, R. N., & Vicente, P. J. (2005). Evidence-based clinical practice guidelines for interdisciplinary rehabilitation of chronic nonmalignant pain syndrome patients. *Pain Practice, 5*(4), 303–315.

Satel, S. L., Kosten, T. R., Schuckit, M. A., & Fischman, M. W. (1993). Should protracted withdrawal from drugs be included in DSM-IV? *American Journal of Psychiatry, 150*(5), 695–704.

Savage, S. R. (2002). Assessment for addiction in pain-treatment settings. *Clinical Journal of Pain, 18*(Suppl 4), S28–S38.

Savage, S. R., Joranson, D. E., Covington, E. C., Schnoll, S. H., Heit, H. A., & Gilson, A. M. (2003). Definitions related to the medical use of opioids. *Journal of Pain and Symptom Management, 26*(1), 655–667.

Savage, S. R., Kirsh, K. L., & Passik, S. D. (2008). Challenges in using opioids to treat pain in persons with substance use disorders. *Addiction Science and Clinical Practice, 4*(2), 4–25.

Schillinger, D., Piette, J., Grumbach, K., Wang, F., Wilson, C., Daher, C., et al. (2003). Closing the loop: Physician communication with diabetic patients who have low health literacy. *Archives of Internal Medicine, 163*(1), 83–90.

Scott, K. M., Hwang, I., Chiu, W.-T., Kessler, R. C., Sampson, N. A., Angermeyer, M., et al. (2010). Chronic physical conditions and their association with first onset of suicidal behavior in the World Mental Health Surveys. *Psychosomatic Medicine, 72*, 712–719.

Sheu, R., Lussier, D., Rosenblum, A., Fong, C., Portenoy, J., Joseph, H., et al. (2008). Prevalence and characteristics of chronic pain in patients admitted to an outpatient drug and alcohol treatment program. *Pain Medicine, 9*(7), 911–917.

Simpson, C. A. (2006). Complementary medicine in chronic pain treatment. *Physical Medicine and Rehabilitation Clinics of North America, 17*(2), 451–472.

Sloman, R., Rosen, G., Rom, M., & Shir, Y. (2005). Nurses' assessment of pain in surgical patients. *Journal of Advanced Nursing, 52*(2), 125–132.

Stalnikowicz, R., Mahamid, R., Kaspi, S., & Brezis, M. (2005). Undertreatment of acute pain in the emergency department: A challenge. *International Journal for Quality in Health Care, 17*(2), 173–176.

Substance Abuse and Mental Health Services Administration. (2007). *Results from the 2006 National Survey on Drug Use and Health: National findings.* NSDUH Series H-32. HHS Publication No. (SMA) 07-4293. Rockville, MD: Author.

Substance Abuse and Mental Health Services Administration. (2008). *Results from the 2007 National Survey on Drug Use and Health: National findings.* NSDUH Series H-34. HHS Publication No. (SMA) 08-43433. Rockville, MD: Author.

Sullivan, M., & Ferrell, B. (2005). Ethical challenges in the management of chronic nonmalignant pain: Negotiating through the cloud of doubt. *Journal of Pain, 6*(1), 2–9.

Tang, N. K., & Crane, C. (2006). Suicidality in chronic pain: A review of the prevalence, risk factors and psychological links. *Psychological Medicine, 36*, 575–586.

Thorn, B. E., Pence, L. B., Ward, L. C., Kilgo, G., Clements, K. L., Cross, T. H., et al. (2007). A randomized clinical trial of targeted cognitive behavioral treatment to reduce catastrophizing in chronic headache sufferers. *Journal of Pain, 8*(12), 938–949.

Trafton, J., Oliva, E. M., Horst, D. A., Minkel, J. D., & Humphreys, K. (2004). Treatment needs associated with pain in substance use disorder patients: Implications for concurrent treatment. *Drug and Alcohol Dependence, 73*, 23–31.

Trescot, A. M., Datta, S., Lee, M., & Hansen, H. (2008). Opioid pharmacology. *Pain Physician, 11*, S133–S153.

Turner, J., Mancl, L., & Aaron, L. (2006). Short- and long-term efficacy of brief cognitive-behavioral therapy for patients with chronic temporomandibular disorder pain: A randomized, controlled trial. *Pain, 121*, 181–194.

U.S. Commission on the Evaluation of Pain. (1987). Appendix C: Summary of the National Study of Chronic Pain Syndrome. *Report of the Commission on the Evaluation of Pain.* Washington, DC: U.S. Government Printing Office.

U.S. Department of Health and Human Services. (2000). *Healthy people 2010: Understanding and improving health.* Stock No. 017-001-001-00-550-9. Washington, DC: U.S. Government Printing Office.

U.S. Food and Drug Administration. (2006). *FDA Public Health Advisory: Combined Use of 5-Hydroxytriptamine Receptor Agonists (Triptans), Selective Serotonin Reuptake Inhibitors (SSRIs) or Selective Serotonin/Norepinenphrine Reuptake Inhibitors (SNRIs) May Result in Life-Threatening Serotonin Syndrome.* Rockville, MD: Center for Drug Evaluation and Research.

Vaillant, G. E. (2003). Natural history of addiction and pathways to recovery. In A. W. Graham, T. K. Schultz, M. F. Mayo-Smith, R. K. Ries, & B. B. Wilford (Eds.), *Principles of addiction medicine* (3rd ed., pp. 3–16). Chevy Chase, MD: American Society of Addiction Medicine.

Van Ameringen, M., Mancini, C., Pipe, B., & Bennett, M. (2004). Antiepileptic drugs in the treatment of anxiety disorders: Role in therapy. *Drugs, 64*(19), 2199–2220.

Vernon, H., Humphreys, K., & Hagino, C. (2007). Chronic mechanical neck pain in adults treated by manual therapy: A systematic review of change scores in randomized clinical trials. *Journal of Manipulative and Physiological Therapeutics, 30*(3), 215–227.

Vitiello, M. V., Rybarczyk, B., Von Korff, M., & Stepanski, E. J. (2009). Cognitive behavioral therapy for insomnia improves sleep and decreases pain in older adults with co-morbid insomnia and osteoarthritis. *Journal of Clinical Sleep Medicine, 5*(4), 355–362.

Volkow, N. D., & Li, T. (2009). Drug addiction: The neurobiology of behavior gone awry. In R. Ries, D. Fiellin, S. Miller, & R. Saitz (Eds.), *Principles of addiction medicine* (4th ed., pp. 3–12). Philadelphia, PA: Lippincott Williams & Wilkins.

Wang, J., & Christo, P. J. (2009). The influence of prescription monitoring programs on chronic pain management. *Pain Physician, 12*(3), 507–515.

Wasan, A. D., Butler, S. F., Budman, S. H., Benoit, C., Fernandez, K., & Jamison, R. N. (2007). Psychiatric history and psychologic adjustment as risk factors for aberrant drug-related behavior among patients with chronic pain. *Clinical Journal of Pain, 23*(4), 307–315.

Weaver, M., & Schnoll, S. (2007). Addiction issues in prescribing opioids for chronic nonmalignant pain. *Journal of Addiction Medicine, 1*(1), 2–10.

Webster, L. R., & Webster, R. M. (2005). Predicting aberrant behaviors in opioid-treated patients: Preliminary validation of the Opioid Risk Tool. *Pain Medicine, 6*(6), 432–442.

Weinstock, J., Barry, D., & Petry, N. M. (2008). Exercise-related activities are associated with positive outcome in contingency management treatment for substance use disorders. *Addictive Behaviors, 33*(8), 1072–1075.

Weissman, D. E., & Haddox, J. D. (1989). Opioid pseudo-addiction—An iatrogenic syndrome. *Pain, 36*(3), 363–366.

Weschules, D. J., Baib, K. T., & Richeimer, S. (2008). Actual and potential drug interactions associated with methadone. *Pain Medicine, 9*(30), 315–344.

Williams, L. S., Jones, W. J., Shen, J., Robinson, R. L., & Kroenke, K. (2004). Outcomes of newly referred neurology outpatients with depression and pain. *Neurology, 63*(4), 674–677.

Wu, S. M., Compton, P., Bolus, R., Schieffer, B., Pham, Q., Baria, A., et al. (2006). The Addiction Behaviors Checklist: Validation of a new clinician-based measure of inappropriate opioid use in chronic pain. *Journal of Pain and Symptom Management, 32*(4), 342–351.

Ziegler, P. (2007). *Treating chronic pain in the shadow of addiction.* Retrieved February 10, 2011, from http://www.emergingsolutionsinpain.com/index.php?option=com_continued&cat=20&Itemid=267

Appendix B—Assessment Tools and Resources

Exhibit B-1 Tools To Assess Pain Level

Tool	Resource
Faces Pain Scale	http://painsourcebook.ca/docs/pps92.html
Numeric Rating Scale	http://www.rnao.org/pda/pain/page4.html
Verbal Rating Scale/Graphic Rating Scale	http://www.rnao.org/pda/pain/page4.html
Visual Analog Scale	http://www.rnao.org/pda/pain/page4.html

Exhibit B-2 Tools To Assess Several Dimensions of Pain

Tool	Resource
Brief Pain Inventory	http://www.mdanderson.org (long and short forms)
McGill Pain Questionnaire	Centre for Evidence Based Physiotherapy http://www.cebp.nl/vault_public/filesystem/?ID=1400 (long form) Center for Gerontology & Health Care Research at Brown University http://www.chcr.brown.edu/pcoc/SHORTMCGILLQUEST.PDF (short form)

Exhibit B-3 Tools To Assess Pain Interference and Functional Capacities

Tool	Resource
Katz Basic Activities of Daily Living Scale	University of Texas School of Nursing at Houston http://son.uth.tmc.edu/coa/FDGN_1/RESOURCES/ADLandIADL.pdf
Pain Disability Index	Pain Balance http://www.painbalance.org/pages/search.aspx?q=pain+disability+index
Roland-Morris Disability Questionnaire	National Primary Care Research and Development Centre, University of Manchester, UK http://www.rmdq.org
WOMAC index	http://www.womac.org

Exhibit B-4 Tools To Screen for Substance Use Disorder

Tool	Resource
Alcohol, Smoking, and Substance Involvement Screening Test	World Health Organization http://www.who.int/substance_abuse/activities/assist/en/index.html
Alcohol Use Disorders Identification Test (AUDIT)	World Health Organization, Department of Mental Health and Substance Dependence http://whqlibdoc.who.int/hq/2001/who_msd_msb_01.6a.pdf
AUDIT-C	Department of Veterans Affairs http://www.hepatitis.va.gov/vahep?page=prtop03-audit_c
CAGE Adapted to Include Drugs (CAGE-AID)	Michigan Quality Improvement Consortium http://www.mqic.org/pdf/CAGE_CAGE_AID_QUESTIONNAIRES.pdf
Drug Abuse Screening Test	Project Cork http://www.projectcork.org/clinical_tools/html/DAST.html
Michigan Alcoholism Screening Test (MAST) (MAST-G for older patients)	National Institute on Alcohol Abuse and Alcoholism http://pubs.niaaa.nih.gov/publications/assesing%20alcohol/instrumentpdfs/42_mast.pdf

Exhibit B-5 Tools To Assess Emotional Distress, Anxiety, Pain-Related Fear, and Depression

Tool	Resource
Beck Depression Inventory (BDI)	It is possible to download the 1961 BDI version, the copyright for which is held by the American Psychological Association rather than Pearson Education. The original BDI is widely available to academic researchers via interlibrary loan, under fair use provisions of international copyright law: Beck, Ward, Mendelson, Mock, & Erbaugh, 1961. Order BDI-II through Pearson at http://pearsonassess.com/HAIWEB/Cultures/en-us/default
Brief Patient Health Questionnaire	Department of Defense/Veterans Health Administration http://www.pdhealth.mil/guidelines/downloads/appendix1.pdf
Center for Epidemiologic Studies Depression Scale	Counselling Resource http://counsellingresource.com/quizzes/cesd/index.html
Geriatric Depression Scale	Stanford University http://www.stanford.edu/~yesavage/GDS.html (long form) http://www.stanford.edu/~yesavage/GDS.english.short.html (short form)
Profile of Chronic Pain: Screen	http://shop.goalistics.com/products/profile-of-chronic-pain
Clinician Administered PTSD Scale	U.S. Department of Commerce National Technical Information Service http://www.ntis.gov/search/product.aspx?ABBR=AVA21105CDRM

Exhibit B-5 Tools To Assess Emotional Distress, Anxiety, Pain-Related Fear, and Depression (continued)

Tool	Resource
Davidson Trauma Scale	Multi-Health systems, Inc. http://www.mhs.com/product.aspx?gr=cli&prod=dts&id=overview
Posttraumatic Diagnostic Scale	Pearson Assessments http://www.pearsonassessments.com
State-Trait Anxiety Inventory	Mind Garden http://www.mindgarden.com/products/staisad.htm
Tampa Scale for Kinesiophobia	http://www.medicalpanels.vic.gov.au/wps/wcm/connect/wsinternet/worksafe/home/forms+and+publications/educational+material/tampa+scale+for+kinesiophobia

Exhibit B-6 Tools To Assess Coping

Tool	Resource
Chronic Pain Acceptance Questionnaire	http://www.somasimple.com/pdf_files/acceptance_pain.pdf (appendix to McCracken, Vowles, & Eccleston, 2004)
Fear-Avoidance Beliefs Questionnaire	WorkSafe Victoria http://www.workcover.vic.gov.au

Appendix C—CFR Sample Consent Form and List of Personal Identifiers

Sample Consent Form

I, _____________________, authorize XYZ Clinic to receive from/disclose to _____________________
(Name of patient/participant)   (Name of person/organization)

for the purpose of ____ the following information
(Need for disclosure)

____.
(Nature of the disclosure)

I understand that my records are protected under the Federal and State Confidentiality Regulations and cannot be disclosed without my written consent unless otherwise provided for in the regulations. I also understand that I may revoke this consent at any time except to the extent that action has been taken in reliance on it and that in any event this consent expires automatically on _________________________________ unless otherwise specified below.
(Date, condition, or event)

Other expiration specifications:____

____   _______________
Signature of patient/participant   Date

____   _______________
Signature of parent/guardian, where required   Date

Individual Identifiers Under the Privacy Rule

The following 18 identifiers of a person or of relatives, employers, or household members of a person must be removed, and the covered entity must not have actual knowledge that the information could be used alone or in combination with other information to identify the individual, for the information to be considered de-identified and not protected health information (PHI):

- Names
- All geographic subdivisions smaller than a State, including county, city, street address, precinct, ZIP Code,* and their equivalent geocodes
- All elements of dates (except year) directly related to an individual: all ages >89 and all elements of dates (including year) indicative of such age (except for an aggregate into a single category of age >90)
- Telephone numbers
- Fax numbers
- Email addresses
- Social Security numbers
- Medical record numbers
- Health-plan beneficiary numbers
- Account numbers
- Certificate and license numbers
- Vehicle identifiers and serial numbers, including license plate numbers
- Medical device identifiers and serial numbers
- Internet universal resource locators (URLs)
- Internet protocol (IP) addresses
- Biometric identifiers including fingerprints and voice prints
- Full-face photographic images and any comparable images
- Any other unique identifying number, characteristic, or code, except that covered identities may, under certain circumstances, assign a code or other means of record identification that allows de-identified information to be re-identified

Source: 45 CFR § 164.514 [b][2][i].

* The first three digits of a ZIP Code are excluded from the PHI list if the geographic unit formed by combining all ZIP Codes with the same first three digits contains >20,000 people.

Appendix D—Resources for Finding Complementary and Alternative Therapy Practitioners

Type of Therapy	Resource
Acupuncture	American Association of Acupuncture and Oriental Medicine (AAAOM) http://www.aaaomonline.org National Certification Commission for Acupuncture and Oriental Medicine (NCCAOM) http://www.nccaom.org
Biofeedback	Association for Applied Psychophysiology and Biofeedback http://www.resourcenter.net/Scripts/4Disapi9.dll/4DCGI/resctr/search.html
Chiropractic	American Chiropractic Association http://www.acatoday.org
Massage	American Massage Therapy Association http://www.amtamassage.org/findamassage/locator.aspx

Appendix E—Field Reviewers

Jane C. Ballantyne, M.D., FRCA
Professor, Anesthesiology and Critical
    Care
Penn Pain Medicine Center
Philadelphia, Pennsylvania

Declan T. Barry, Ph.D.
Associate Research Scientist
Yale University School of Medicine
New Haven, Connecticut

Sharon M. Freeman, Ph.D., M.S.N.,
    PMHCNS-BC, CARN-AP
CEO
Center for Brief Therapy, P.C.
Fort Wayne, Indiana

Cynthia M. A. Geppert, M.D.,
    Ph.D., M.P.H., FAPM, DAAPM
Chief
Consultation Psychiatry & Ethics
New Mexico Veterans Affairs Health
    Care System
Associate Professor of Psychiatry
Director of Ethics Education
University of New Mexico School of
    Medicine
Albuquerque, New Mexico

Patricia M. Good
Chief (retired)
Liaison and Policy Section
Office of Diversion Control
Drug Enforcement Administration
Arlington, Virginia

Douglas Gourlay, M.D., M.Sc.,
    FRCP(C), FASAM
Director
Pain and Chemical Dependency
    Division
Wasser Pain Centre
Mount Sinai Hospital
Centre for Addiction and Mental
    Health
Toronto, Canada

Howard A. Heit, M.D., FACP,
    FASAM
Assistant Clinical Professor of
    Medicine
Georgetown School of Medicine
Fairfax, Virginia

Nathaniel Katz, M.D., M.S.
Adjunct Assistant Professor of
    Anesthesia
Tufts University School of Medicine
President & CEO
Analgesic Research
Needham, Massachusetts

Kenneth L. Kirsh, Ph.D.
Assistant Professor
Pharmacy Practice and Science
University of Kentucky
Lexington, Kentucky

Paul Kreis, M.D.
Professor, Medical Director
Division of Pain Medicine
University of California, Davis
Sacramento, California

Pamela A. Pavilonis, M.B.A., M.S., N.D.
Naturopathic Physician
West Linn, Oregon

Seddon R. Savage, M.D., M.S.
Director
Dartmouth Center on Addiction Recovery
    and Education
Hanover, New Hampshire
Pain Consultant
Manchester VA Medical Center
Manchester, New Hampshire

Randy Seewald, M.D.
Medical Director
Methadone Maintenance Treatment Program
Beth Israel Medical Center
New York, New York

Barbara St. Marie, ANP, GNP
Nurse Practitioner
Burnsville, Minnesota

Mark D. Sullivan, M.D., Ph.D.
Professor of Psychiatry and Behavioral
    Sciences
University of Washington
Seattle, Washington

Andrea Trescot, M.D.
Professor
Department of Anesthesia and Pain Medicine
Director
Pain Fellowship Program
University of Washington
Seattle, Washington

**Norman Wetterau, M.D., FAAFM,
    FASAM**
New York, New York

Penelope P. Ziegler, M.D., FASAM
Medical Director
Virginia Health Practitioners' Monitoring
    Program
Richmond, Virginia

Appendix F— Acknowledgments

Numerous people contributed to the development of this TIP, including the TIP Consensus Panel (see p. ix) and TIP field reviewers (see Appendix E).

This publication was produced under the Knowledge Application Program (KAP), a Joint Venture of JBS International, Inc. (JBS), and The CDM Group, Inc., and for the Substance Abuse and Mental Health Services Administration, Center for Substance Abuse Treatment.

Lynne MacArthur, M.A., A.M.L.S., served as the JBS KAP Executive Project Co-Director, and Barbara Fink, RN, M.P.H., served as the JBS KAP Managing Project Co-Director. Other JBS KAP personnel included Candace Baker, M.S.W., MAC, Deputy Director for Product Development; Kris Rusch, M.A., Senior Writer; Cathy Baker, M.Ed., Senior Writer; Martha Horn, Ph.D., Senior Writer; Frances Nebesky, M.A., Quality Assurance Editor; Wendy Caron, Quality Assurance Manager; Erin P. Doherty, Copy Editor; and Erin Sandor, Graphic Designer.

Index

A

CSAT TIPs and Publications Based on TIPs

What Is a TIP?

Treatment Improvement Protocols (TIPs) are the products of a systematic and innovative process that brings together clinicians, researchers, program managers, policymakers, and other Federal and non-Federal experts to reach consensus on state-of-the-art treatment practices. TIPs are developed under CSAT's Knowledge Application Program to improve the treatment capabilities of the Nation's alcohol and drug abuse treatment service system.

What Is a Quick Guide?

A Quick Guide clearly and concisely presents the primary information from a TIP in a pocket-sized booklet. Each Quick Guide is divided into sections to help readers quickly locate relevant material. Some contain glossaries of terms or lists of resources. Page numbers from the original TIP are referenced so providers can refer back to the source document for more information.

What Are KAP Keys?

Also based on TIPs, KAP Keys are handy, durable tools. Keys may include assessment or screening instruments, checklists, and summaries of treatment phases. Printed on coated paper, each KAP Keys set is fastened together with a key ring and can be kept within a treatment provider's reach and consulted frequently. The Keys allow you—the busy clinician or program administrator—to locate information easily and to use this information to enhance treatment services.

Ordering Information

Publications may be ordered for free at http://store.samhsa.gov. To order over the phone, please call 1-877-SAMHSA-7 (1-877-726-4727) (English and Español). Most publications can also be downloaded at http://www.kap.samhsa.gov.

TIP 1 **State Methadone Treatment Guidelines—**_Replaced by TIP 43_

TIP 2* **Pregnant, Substance-Using Women—**BKD107

TIP 3 **Screening and Assessment of Alcohol- and Other Drug-Abusing Adolescents—**_Replaced by TIP 31_

TIP 4 **Guidelines for the Treatment of Alcohol- and Other Drug-Abusing Adolescents—**_Replaced by TIP 32_

TIP 5 **Improving Treatment for Drug-Exposed Infants—**BKD110

TIP 6* **Screening for Infectious Diseases Among Substance Abusers—**BKD131
Quick Guide for Clinicians QGCT06
KAP Keys for Clinicians KAPT06

TIP 7 **Screening and Assessment for Alcohol and Other Drug Abuse Among Adults in the Criminal Justice System—**_Replaced by TIP 44_

TIP 8 **Intensive Outpatient Treatment for Alcohol and Other Drug Abuse—**_Replaced by TIPs 46 and 47_

TIP 9 **Assessment and Treatment of Patients With Coexisting Mental Illness and Alcohol and Other Drug Abuse—**_Replaced by TIP 42_

TIP 10 **Assessment and Treatment of Cocaine-Abusing Methadone-Maintained Patients—**_Replaced by TIP 43_

TIP 11* **Simple Screening Instruments for Outreach for Alcohol and Other Drug Abuse and Infectious Diseases—**BKD143
Quick Guide for Clinicians QGCT11
KAP Keys for Clinicians KAPT11

TIP 12 **Combining Substance Abuse Treatment With Intermediate Sanctions for Adults in the Criminal Justice System—**_Replaced by TIP 44_

TIP 13 **Role and Current Status of Patient Placement Criteria in the Treatment of Substance Use Disorders—**BKD161
Quick Guide for Clinicians QGCT13
Quick Guide for Administrators QGAT13
KAP Keys for Clinicians KAPT13

*Under revision

*Under revision

*Under revision